Combinatorial Physics

SERIES ON KNOTS AND EVERYTHING

Editor-in-charge: Louis H. Kauffman

Published:

Vol. 1:   Knots and Physics
          L. H. Kauffman

Vol. 2:   How Surfaces Intersect in Space
          J. S. Carter

Vol. 3:   Quantum Topology
          edited by L. H. Kauffman & R. A. Baadhio

Vol. 4:   Gauge Fields, Knots and Gravity
          J. Baez & J. P. Muniain

Vol. 6:   Knots and Applications
          edited by L. H. Kauffman

Vol. 7:   Random Knotting and Linking
          edited by K. C. Millett & D. W. Sumners

Vol. 10:   Nonstandard Logics and Nonstandard Metrics in Physics
          W. M. Honig

Forthcoming:

Vol. 5:   Gems, Computers and Attractors for 3-Manifolds
          S. Lins

Vol. 8:   Symmetric Bends: How to Join Two Lengths of Cord
          R. E. Miles

KoE Series on Knots and Everything — Vol. 9

COMBINATORIAL
PHYSICS

T Bastin
C W Kilmister

Singapore • New Jersey • London • Hong Kong

Published by
World Scientific Publishing Co. Pte. Ltd.
P O Box 128, Farrer Road, Singapore 9128
USA office: Suite 1B, 1060 Main Street, River Edge, NJ 07661
UK office: 57 Shelton Street, Covent Garden, London WC2H 9HE

COMBINATORIAL PHYSICS

Copyright © 1995 by World Scientific Publishing Co. Pte. Ltd.

All rights reserved. This book, or parts thereof, may not be reproduced in any form or by any means, electronic or mechanical, including photocopying, recording or any information storage and retrieval system now known or to be invented, without written permission from the Publisher.

For photocopying of material in this volume, please pay a copying fee through the Copyright Clearance Center, Inc., 222 Rosewood Drive, Danvers, Massachusetts 01923, USA.

ISBN 981-02-2212-2

This book is printed on acid-free paper.

Printed in Singapore by Uto-Print

Preface

It is nearly fifty years since the authors of this book began a collaboration based on their common interest in the foundations of physics. During that time others have made very major contributions. The fruits of this cooperative enterprise, particularly of the later part of it, are set out here. One problem has led to another. Over that time it became clear that such existing preconceptions as the space and time continua formed an inadequate basis for a physics which has to incorporate a quantum world of discrete character. Here we argue that the impossibility of reconciliation between continuous and discrete starting points means that we must start from the discrete or combinatorial position. Otherwise the quantum theory will remain with confusion and muddle at its centre.

If intuitive clarity is to come from the combinatorial approach it turns out to be hard won because continuum ideas are so deeply embedded in orthodox physics. It has been possible to travel only part of the way but that has been far enough to reveal another positive aspect of the approach. Discreteness is intimately related, according to our theory, to the existence of *scale-constants* —those dimensionless constants commonly thought by physicists to be of some fundamental significance. We should therefore be able to calculate these. Here we are able to detail the calculation of one, the fine-structure constant, which Paul Dirac emphasized for much of his life as an outstanding problem in the completion of quantum electrodynamics. Our value agrees with the experimentally determined one to better than one part in 10^5. The calculation of dimensionless constants will bring to mind the name of Eddington, and although his work was the original cause of the authors' meeting, and although they agree with him in seeking a combinatorial origin for these constants, their mathematical method, and certainly their calculations of the values of those, have nothing in common with his.

This is a book about physics but philosophers will find that some issues — once their province — which they thought dead and decently buried, are resurrected to new life here. Notable amomg these is the place of mind. The great originators of the quantum theory knew that the action of the mind (or the "observer") had to be part of the theoretical structure, but this development has been aborted. In combinatorial theory there is no escape from the issue. In somewhat the same way, since computing is essentially combinatorial, people in and around computer science may find our representation of physics more natural to them than it is to some orthodox physicists. None the less it is primarily the physics community we seek to inspire to carry further a project of which this is the beginning.

Contents

Preface		v
Chapter 1	Introduction and Summary of Chapters	1
2	Space	7
3	Complementarity and All That	17
4	The Simple Case for a Combinatorial Physics	25
5	A Hierarchical Model — Some Introductory Arguments	33
6	A Hierarchical Combinatorial Model — Full Treatment	57
7	Scattering and Coupling Constants	91
8	Quantum Numbers and the Particle	121
9	Toward the Continuum	135
10	Objectivity and Subjectivity — Some 'isms'.	153
References		165
Name Index		169
Subject Index		173

CHAPTER 1

Introduction and Summary of Chapters

This book is an essay in the conceptual foundations of physics. Its purpose is to introduce what we shall call a combinatorial approach. Its method will be to view physical theory under the aspect of a particular point of view which is combinatorial in character. In the course of the book questions will be asked and discussed which may have a long history, but which are not seen as live issues at the present because of the special philosophical stance of present-day physicists. However for combinatorial physics they are very much alive.

The idea which underlies combinatorial physics is that of *process*. The most fundamental knowledge that we can have is of step-by-step unfolding of things; so in a sequence. This is the kind of knowledge we have of quantum processes, and that fact becomes specially evident in the experimental techniques of high-energy physics.

The contrasting view, which has been the main guide of physics in the past, is of a background of physical things whose spatial relationships are on an equal footing in the sense that it can only be an accident in what order we happen across them. To put it another way, when the sequence or process is fundamental then we have to specify the steps by which we get to a particular point, whereas in the conventional picture we imagine we are free to move about in the space in which phenomena occur, without its being necessary to be explicit about it.

Suggestions for fundamental revision of the conceptual framework of physics are unlikely to engage the attention of physicists unless they bring about major improvements in the technical understanding of physics — particularly in new explanations and calculations of experimental results. The strong position of the combinatorial theory is that it has been used to deduce some experimental quantities which have not been deduced from more conventional theory. The closeness to the experimental values of these deductions makes it very unlikely that their success is fortuitous. Moreover it would be agreed on all sides that the experimental quantities in question play a sufficiently important part in physics to warrant attention's being paid to any theory which claims to calculate them.

$$* \quad * \quad * \quad * \quad *$$

We have to be more explicit about what we mean by a combinatorial physical theory. Combinatorial physics is physics in which the mathematical relations are combinatorial, and combinatorial mathematics is mathematics in which we study the ways in which symbols are combined. The term 'combinatorial' is often defined ostensively by giving examples. Ryser[1] gives the following examples: interest in magic squares, study of permutations and (indeed) of combinations, finite projective geometry and problems connected with covering of spaces with arrangements of shapes in circumstances in which the space is divided into a finite number of sections and there is a convention which enables us to decide unambiguously whether or not a given section is covered. Such conventions exist, for example, for the use of chess boards and most other gaming boards, and the conventions enable us to decide all questions of the relation of combinatorial structures to physically defined spaces. Ryser sums up his exemplification thus: "Combinatorial mathematics cuts across the many subdivisions of mathematics, and this makes a formal definition difficult. But by and large it is concerned with the study of the arrangements of elements into sets."

The first person to have seen a very profound difference between the combinatorial approach and the rest of formalised thinking was Leibniz. The term 'combinatorial', used in this context, originates with Leibniz's "Dissertate de arte combinatoria", and according to Polya[2] the originator saw that difference in a very striking form. ... "Leibniz planned more and more applications of his combinatorial art or 'combinatorics': to coding and decoding, to games, to mortality tables, to the combination of observations. Also he widened more and more the scope of the subject. He sometimes regards combinatorics as one half of a general Art of Invention; this half should deal with synthesis, whereas the other half should deal with analysis. Combinatorics should deal, he says in another passage, with the same and the different, the similar and the dissimilar, the absolute and the relative, whereas ordinary mathematics deals with the one and the many, the great and the small, the whole and the part."

This insight of Leibniz seems borne out by what is happening in formal areas of thinking at the present time. The general connection between combinatorial mathematics and the impulse behind computing science has been widely remarked. There is more to it than that computer programmes are constructed in formal steps with information stored in (binary) patterns. It is the connectivity of the computer programme which is its vital characteristic, and which is an aspect of the ars combinatoria. Weyl[3] says, "Modern computing machines translate our insight into the combinatorial structure of mathematics into practice by mechanical and electronic devices." When one is subjecting oneself to the discipline of the computer programming, which in its essentials is arranging the mathematical operations in sequential order so that each brings the next appropriate one in its train without intervention on the part of the mathematician, then Weyl is saying that one has somehow, through exercise of that discipline, displayed that insight.

Arguing this way, one comes to feel that in spite of all the modern sophistication about the impact of computing on the foundations of mathematics, not enough attention has been paid to the vast difference between the classical idea of the mathematician and that required by the computer revolution. This difference shows up clearly when it comes to making mathematical models. The former makes up his mind about how to get from point to point in a mathematical structure on the basis of some purpose that he imagines to be directing his efforts, whereas, by contrast a computer model is meant to have these instructions incorporated in the program.

In contemporary thinking there are a variety of essential principles which are welded together in a mathematical framework which we call quantum mechanics. Uncertainty, exclusion, complementarity and the whole theory of wave-functions and differential operators and eigenstates are some of these. From this point of view anyone who undertakes to reformulate the foundations of quantum physics will be expected to replicate this same mathematical structure because that simply is what quantum theory is. At this early stage we should warn the reader not to have this unquestioning expectation.

Our case is that all the results of current quantum theory are combinatorial in origin and — at any rate in principle — we should expect to obtain them all by continuation of our combinatoric method. Moreover we argue that all the principles which are normally thought to be characteristic of the quantum theory take their familiar form because of the need to reconcile the discrete character of quantum events with the classical theory-language.

Plan of the book by chapters

CHAPTER 1. INTRODUCTION
The book is an essay in the foundations of physics; it presents a combinatorial approach; ideas of *process* fit with a combinatorial approach; quantum physics is naturally combinatorial and high energy physics is evidently concerned with process. Definition of 'combinatorial'; the history of the concept takes us back to the bifurcation in thinking at the time of Newton and Leibniz; combinatorial models and computing methods closely related.

CHAPTER 2. SPACE
Theory-language defined to make explicit the dependence of modern physics on Newtonian concepts, and to make it possible to discuss limits to their validity; Leibniz' relational, as opposed to absolute, space discussed; the combinatorial aspect of the monads.

CHAPTER 3. COMPLEMENTARITY AND ALL THAT
Bohr's attempt to save the quantum theory by deducing the wave-particle duality, and thence the formal structure of the theory, from a more general principle (complementarity) examined: the view of complementarity as a philosophical gloss on a theory which stands up in its own right shown to misrepresent Bohr: Bohr's argument rejected — leaving the quantum theory still incomprehensible.

CHAPTER 4. THE SIMPLE CASE FOR A COMBINATORIAL PHYSICS
Physics not scale-invariant; it depends on some numbers which come from somewhere outside to provide absolute scales; the classical kind of measurement cannot in the nature of the case provide them; measurement is counting; the coupling constants are the prima-facie candidates; this was Eddington's conjecture; the question is not *whether* we find combinatorial values for these constants, but how we do so; current physics puts the values in *ad hoc*.

CHAPTER 5. A HIERARCHICAL MODEL — SOME INTRODUCTORY ARGUMENTS
The combinatorial model used is hierarchical; the algorithm relating the levels is due to Parker–Rhodes; the construction is presented in several ways each stressing a particular connection with physics: (1) similarity of position, (2) the original combinatorial hierarchy, (3) counter firing, (4) limited recall, (5) self-organization, (6) program universe.

CHAPTER 6. A HIERARCHICAL COMBINATORIAL MODEL — FULL TREATMENT

The elementary process expressed algebraically and interpreted as decision whether to incorporate a presented element as new; new elements labelled; the need for labelling to be consistent gives central importance to discriminately closed subsets; any function which can assign labels equivalent to one which represents process; process defined as always using the smallest possible extension at each step that is allowed by the previous labelling; representation of functions by arrays; representation of arrays by matrices and strings familiar from the simpler treatments of Chapter 5; summary of the argument.

CHAPTER 7. SCATTERING AND COUPLING CONSTANTS

The primary contact with experiment in quantum physics comes through counting in scattering processes; coupling constants are ratios of counts which specify the basic interactions; this outline picture has to be modified to get the experimental values; history of attempts to calculate the fine-structure constant reviewed; explanations of and calculations of the non-integral part due to McGoveran and to Kilmister given. The latter follows better from the principles of construction of the hierarchy algebra of Chapter 6.

CHAPTER 8. QUANTUM NUMBERS AND THE PARTICLE

Comments provided on high energy physics and the particle/quantum number concept from the standpoint regarding the basic interactions of Chapter 7; the particle is the conceptual carrier of a set of quantum numbers; the view of the particle as a Newtonian object with modifications is flawed; an alternative basis for the classification of the quantum numbers due to Noyes is described; it is compared with the Standard Model.

CHAPTER 9. TOWARDS THE CONTINUUM

We have no representation of physical space, let alone the continuum; the conventional understanding of dimensionality replaced by a 3D argument based on the hierarchy algebra; the finite velocity of light necessarily follows from the pure-number fine-structure constant; it leads to a very primitive form of relativity; this is developed; the quadratic forms which appear in the Lorentz transformation as well as in Pythagoras' theorem are discussed; measurement is defined.

CHAPTER 10. OBJECTIVITY AND SUBJECTIVITY — SOME "ISMS"

The philosophical position of the book is assessed to see how it fits with some familiar positions — mostly ending in "ism": subjectivism; realism; the anthropic principle; constructivism; reductionism; the critical philosophy; positivism; operationalism; particles.

CHAPTER 2

Space

The most important consequence of the change to combinatorial mathematics in physics arises over the representation of physical space. What we abandon is the automatic freedom to consider a problem from several points of view and involving several mathematical techniques which have no coherence but which we take as simultaneously relevant because we ascribe them all to the same point of space. The combinatorial approach has to construct a process which is equivalent as far as possible to the classical putting together of results to give a sort of composite picture of what is happening at a point of space. In this book we shall only be able to take the first steps in this programme — just enough to show what is needed. There is much that comes first.

If one wishes to replace spatial relationships as the primary data by sequential development, then one should first look at the reasons for the profound hold that the former way exerts. It postulates a continuum background of space within which things have positions, and the changes in those positions give rise to a second continuum of time. These continua are imagined as perfectly smooth, perfectly homogeneous, infinitely divisible. They are modelled mathematically by the continuum of all real numbers as that was formalized by Dedekind and Cantor between 1870 and 1880. The irreducibly simplest entities in these continua of space and time are particles — idealized as single points whose position in space changes continuously with time — and fields. Fields are spread through space and have at each spatial point a magnitude which again varies smoothly both with time and with changes of the point at which the field is specified. Upon this basis, a scheme of dynamical concepts was elaborated which seemed to provide a complete description of the motion of things in the space continuum. The work of Newton was the cornerstone of this edifice. Its subsequent elegance (epitomized by Thomson and Tait's "Natural Philosophy") compelled a feeling of universality: one felt one was in possession of the means to describe reality direct. Indeed this scheme became for the physicist the mathematical elaboration of commonsense and the automatic vehicle for his thought.

The set of concepts that makes up the scheme has an interlocking and closed character which makes it unique in the history of thought. Other disciplines have aspired to this cohesion, but have only very partially been able to achieve it. Through its long elaboration classical mechanics has become like a language which we learn to use rather than like an exploration which may turn out right or may turn out wrong, though of course it was like that once. When we learn classical mechanics we find that the definitions of the concepts chase themselves round in circles. The mass of a body is given a numerical value through the behaviour of that body under a given applied force. The force is specified numerically by the acceleration of the body, but only if we already know the mass, and so on. At the time of Newton, there was no unanimity on the use of the concepts, and if one were asked to give a definition of one of them, one would refer to some experimental situation which made the use one was proposing plausible: one would have to argue. Now things are interestingly different, for the set of interlocking concepts defines its own appropriate application, and therefore cannot conflict with experiment.

It is easy to give an example. It is now known that the spiral nebulae rotate like a solid body, so that parts of the nebula at different distances out on a radius stay on that radius. Not merely that, but there may be no simple relation between the rotation and the swept back look: in certain cases the arms may advance point first. After those shocks one is almost disappointed to find that the thing does rotate in the plane in which its arms lie, like a catherine wheel. Yet, one would have sworn, if ever there were an object whose rough dynamics was plain for all to see, it would have been the spiral nebula — a loose aggregate of matter having some angular momentum, and with some radial motion shown by distribution of material along the arms, but with the arms appearing swept back because of greater angular velocities near the centre.

The fascinating subject of galactic dynamics is, however, not our topic now: we are using it only to make a methodological point. This is that no amount of evidence will induce us to doubt the asumption that the universe out there would seem as it does to us here, and exhibit the same mechanics, if we could be transported there. We know we have to give up that presumption as we approach the atomic scale: that we retain it in the face of possible contrary evidence at the increasingly large scale is something between a guiding principle and a prejudice. It should be re-evaluated in the light of the information we actually have each time we use it. The strongest case for the principle or prejudice is that galaxies have similar form over large variations in their red-shift, and so can be regarded to a first approximation as invariant units. The spiral galaxies are the largest structures that we encounter on our way to the 'limit of the observable universe'. There, the extrapolation principle compels us to entertain the idea of a universe which goes on for ever and ever, which is unsatisfactory when we need to operate theoretically with the "observable universe".

It is a strange fact that this immunity of the deductive language of classical physics to experimental check has received little or no attention in contemporary writing on the philosophy of physics. The authors[1] coined the term *theory-language* to refer to a theory which had reached this stage of development. They tried to use the 'logic of facts' of Wittgenstein's Tractatus Logico-philosophicus, to describe it: "... a physical theory consists of propositions which may be thoughts, sentences written or spoken, or manipulations with bits of the physical world ... the experimental thing has meaning only as part of a theory. The theory may have different degrees of complexity, and there will be experimental procedures corresponding to each degree. Thus the theory is a kind of language, but experiments in the theory are the same language. One cannot use experiments in a complex language to criticise a simpler theoretical language."

What has to be replaced is the 'platonic receptacle' view of space: space is what holds whatever we care to put into it. Of course the receptacle view is deep in our thinking: it seems to be a way of operating physical theory by projecting our most immediate sensuous knowledge of the world onto each theoretical statement — it is almost as though we need to picture what bodily actions we should take to correspond to it, before we can understand it. If we go back to our example of the spiral nebula, it is as though we transport ourselves in the imagination to that place and think what would happen to us in order to formulate the theoretical description. Our attempt to formalize the classical theory-language was essentially to replace this sensuous correspondence. We have to remember that when we speak of replacing the continuum we are replacing this too.

If we associate the 'classical theory-language' predominantly with Newton, it is not surprising to find the alternative associated predominantly with that other giant — his contemporary — Leibniz. The differences in conception of the two men to be seen in their respective innovations which became two versions of the calculus, were pointers to a greater difference. We have already associated Leibniz' differing view with the combinatorial aspect of his thought. This seems to have been intrinsic to his view of the world in rather the way we have postulated with our view of the centrality of process. If, for ease of expression, we let a computing model stand for combinatorial process, then we have in the nature of the case to view every process from some preferred viewpoint. This viewpoint is whatever is at the moment in the central processor together with its connections. (The possibility of parallel or concurrent computing is not relevant to this argument and arises at a less fundamental stage: If there are several computers going, then insofar as we can speak of them as one model they are one computing system, even though the connections between them may be randomized). Leibniz faces this profound problem of escaping from the preferred starting point or reference point and enables us to see it as of great generality. For us, it will be an important piece to fit into the theory.

For the great majority of physicists, there is no problem in understanding the success of the theory-language of classical physics: they do not have to worry how it comes about that the theory-language describes the world so automatically because there is no distinction between it and reality. In a way we are suffering from the enormous success of classical physics. Over three centuries its concepts have penetrated our common language about the world almost completely, and the resulting amalgam of theory and experience seen under the aspect of that theory has become commonsense; is identified with reality; and has the corresponding incorrigibility.

There is a trivial form of closed language which is used to describe the world, which is quite different from the theory-language. This could well be called the 'self-authenticating language'. It is commonest in religious literalism or fundamentalism where the language is contrived to bar the way of serious enquiry by going round in circles. "You should pay unique attention to what it says in this book because it is the Word of God. You will find out about God in the book." The difference in the case of the theory-language is that wide ranges of our experience of the world really conform to it.

It is a little puzzling to say precisely in what this correspondence with reality consists, but it seems to be connected with the prevalence of the dynamics of the particle. The history of physics from Galileo onwards has been the progressive displacement by the dynamics of the particle of all other forms of explanation. Descartes' vortices failed because it proved not to be possible to calculate with them, and 'calculate' increasingly came to mean 'calculate using the dynamics of the particle'. It is true that the great nineteenth century inventions of electromagnetism and fluid mechanics, which depend upon the theory of fluid rotation, could be seen as a vindication of Descartes' vortices, but really this would be to stretch a point, for the mode of action which Descartes imagined was not of particles as was the case with the nineteenth century theory. All fields are specified in terms of the behaviour of ideal 'test-particles': we don't understand the mechanics of rigid bodies and such things as gyroscopic motion until we have satisfied ourselves that the special concepts which they introduce can be made explicable in terms of particle dynamics:

fluid mechanics is in no doubt that one must start with the 'experience' of the particle moving with the fluid, though it has two ways of interpreting that requirement.

One reason why we tried to get to a formal representation of the theory-language was to make it possible to ask the question 'what happens when the theory-language really fails to work?' The task was to try to find some mathematical specification of the theory-language which would, as it were, let us lift it off its application and carry it around to see where it would fit; where it would not fit; and why. There is an obvious logical difficulty, if not actually an absurdity, in asking for a theory which accounts for, or even describes, the limits of applicability of itself. You must have something like a grid on which it is mapped to do that, and classical mechanics does not envisage anything of that sort. Both the great innovatory theories of this century — the quantum theory and relativity — found themselves up against this difficulty, and for that reason early forms of both theories took the form of arbitrarily imposed restrictions or alterations to the existing theory which came into effect at certain scales. This made the failure to fit of the theory look like an additional piece of mechanics of the type that the theory dealt in, when the true origins of the limits in each case were really quite different. The incorporation of the constant h into the mathematics of the quantum theory, and the Lorentz contraction, respectively, were the points of application of this method in the two theories. In these ways it was possible to put off the recognition that familiar universal dynamical concepts no longer applied at all.

Subsequently the theories became more sophisticated, but they haven't gone all the way yet. The very ways of speaking about non-locality still show a backward-looking stance. One expects normal locality and takes to special devices when it fails, instead of seeing a more general form of connectivity as the norm. Very similar remarks apply to the quantum limits. Discussion of how one can have a dynamics without a classical background space and time when one adopts a combinatorial point of view will occupy a large part of this book, and we can stay at the level of general comment here. At that level it may be worth pointing out one principle which makes limits to the applicability of familiar dynamics easier to understand. One is accustomed to have a description of the world which is derived from our everyday laboratory-scale experience and to extrapolate it to different magnitudes. The technique and its very language suggest nothing which is not scale-invariant, and therefore we are puzzled to find it failing to work at particular scales, and have no way of describing that failure. How are we to understand limits to extrapolation? One piece of the puzzle arises from the way we take it for granted that there are an indefinite number of quite distinct ways of making the observations upon which our physical descriptions are based, and that these give consistent information. However, as we extrapolate further in both the small and the large scales, this presumption gets further and further from being valid. So far from having an indefinite number of cross-referencing methods of observing, we are reduced to very few, and ultimately to one. Notoriously, all our information about the universe on the largest scale depends on an interpretation of the red shift of spectra, while at the quantum end we are using the very particles we are investigating to do the investigating.

A quite insightful way of looking at what is happening is to say the theory and observation are losing their separability. By 'separability' we mean the situation, usually taken for granted, of a body of experimental results which is then rationalized or simplified to a greater or lesser extent by one of a number of possible theories. The choice of theory is influenced by, amongst other things, the extent of the rationalization produced. Here by contrast one gets towards a state where one would not know how to present the results without presupposing one particular theory. None of this discussion is in any way to impugn the validity of the observational work for what it is, nor to cast doubt on the care and sophistication with which deductions from it are made. It is only that current thinking allows for only one category of experimental knowledge, whereas we are confronted with something like a sliding scale from a situation where consistency is assured to one where it has no meaning. The very notion of observation has undergone a change in logical character. If we recognize that the experimental evidence and the theory can no longer be separated it is easier to understand the limit to extrapolation. It is usually true that adequate critical standards are applied to interpretations of experimental evidence near the extrapolation limits, yet the absence of a way of speaking which gives the experimental evidence its proper status cannot but cause confusion and possibly misdirection of effort. All that talk of "the first three minutes" will certainly mislead people into taking the time language literally, and it is by no means certain that none of them will be professionals.

The world view of Leibniz — seen in relation to the Newtonian legacy of modern atomism — is so well described by Whitehead[2] that a quite long quotation from him may be in order. "He" (Newton) "held the most simple-minded version of the Lucretian doctrine of the void, the most simple-minded version of the Lucretian doctrine of material atoms, and the most simple-minded version of Law imposed by Divine decree. His only approach to any metaphysical penetration arises from his way of connecting the Void with the sensorium of the Divine Nature. ... The monads of Leibniz constitute another version of the atomic doctrine of the Universe. ... Leibniz was acutely conscious of ... the problem of the criticism of knowledge. Thus he approached the problem of cosmology from the subjective side, whereas Lucretius and Newton approach it from the objective point of view. They implicitly ask the question, What does the world of atoms look like to an intellect surveying it? ... But Leibniz answered another question. He explained what it must be like to be an atom. Lucretius tells us what an atom looks like to others, and Leibniz tells us how an atom is feeling about itself. ... Leibniz wrestles with a difficulty which infects modern cosmologies, a difficulty which Plato, Aristotle, Lucretius, and Newton entirely ignore. ... The modern outlook arises from the slow influence of Aristotle's logic, during a period of two thousand years. ... keeping to the point of view derived from Aristotle's logic, if we ask for a complete account of a real particular thing in the physical world, the adequate answer is expressed in terms of a set of ... abstract characteristics, which are united into an individual togetherness which is the real thing in question."

For Leibniz, this way of handling the particle concept is the only one possible. There never is a case where (as in the modern correspondence principle) it gradually becomes legitimate to go over into a classical mode of thought. We shall find this trenchant criticism having an effect through this book like Coleridge's fearful fiend, which 'doth close behind us tread', though we hope we shall summon up courage to turn round and look at it when we need to.

Many physicists will be puzzled by Whitehead's comments on the subjective approach of Leibniz-feeling that there surely could be no place for such fanciful questions as how the atom sees itself. However the conventional alternative has this at least equally strange component that — as we argued above — it requires us to incorporate our sensuous experience whenever we try to describe the behaviour of a system of particles in a physical space. Whitehead is really drawing attention to Leibniz' doctrine on matter: an object is defined by the bundle of attributes which it has. Such a definition should be congenial to particle physics. But it does raise a problem — that of continuing identity — which is neatly dodged in the Newtonian synthesis by appealing to a continuity in temporal and spatial positions. No such appeal is open to Leibniz, (nor, incidentally to us) and this leads him on to the doctrine of monads.

Whitehead goes on to say of Leibniz's position that it "is beautifully simple. But it entirely leaves out of account the interconnections between real things. Each substantial thing is thus conceived as complete in itself, without reference to any other substantial thing. Such an account of the ultimate atoms or of the ultimate monads, or of the ultimate objects enjoying experience, renders an interconnected world of real individuals unintelligible. The universe is shivered into a multitude of disconnected substantial things, each thing in its own way exemplifying its private bundle of abstract characters which have found a common home in its own substantial individuality. But substantial thing cannot call unto substantial thing." This is the famous problem of the windowless monads. The system itself is thus necessarily incomplete, and Leibniz recognized this lack by breaking the system to the extent of allowing one supreme monad to have windows toward the rest. As far as his writing for the world of his contemporaries was concerned he could make a virtue of this necessity by identifying the supreme monad with the agglomeration of ideas that made up the then current perception of God. Whitehead comments "such has been the long slow influence of Aristotelian logic upon cosmological theory. Leibniz was the first, and by far the greatest philosopher who both accepted the modern" (outlook) "and frankly faced its difficulty".

The picture presented in the arcane writings of Leibniz which only became available at the beginning of our century is rather different. It is clearer from that picture how the monadology fitted Leibniz's concern with patterns of combinations of things. From it we get a vision of the monadology as the abstract form from which one can see what advantages and disadvantages are bound to attend any theory which follows a combinatorial line. Referring to this work, Russell[3] concludes: "What I, for my part, think best in his theory of monads is his two kinds of space, one subjective, in the perceptions of each monad, and one objective, consisting of the points of view of the various monads. This, I believe, is still useful in relating perception to physics."

So we shall find it. We shall postulate a background manifold of entities about which, to put it in Leibniz' terms, the monad obtains information which is prior to the construction of a common space. Then in getting to the public space we avoid Leibniz's difficulty about the windows by use of a principle of indifference in which the only formulations which interest us as physicists are those which would remain true if we gave equal prior probability to all the things which could happen. Physical laws are then automatically things upon which all monads would agree. In this way we shall get a view of physics as a very special case of the totality of our possible experience in which objectivity is reached by employing a statistical technique (albeit a logically very different kind of statistical technique from that used in quantum theory).

We appear at first sight to be departing from Leibniz' thought. He envisages what does not exist as struggling to exist; not all possibles can exist because they may not be 'compossible' and he sees the existent as the state which is compatible with more things existing than any other. Indeed it sounds very much as though we are appealing to the principle sometimes known as that of 'plenitude', according to which everything which could happen could be relied upon to happen given enough time, and this principle Leibniz most vehemently dissociated himself from, where it is to be found in the writings of Descartes and others. However it must be remembered that we are only taking on physics, whereas Leibniz was seeing a much wider sweep of country. It will be shown later that though our numerical results follow from our principle of indifference, which has this resemblance to the principle of plenitude, it is essential as part of the technical detail of the theory that the principle only applies as a special case. It is what happens when chaos reigns, but more intelligible structure will in general exist in the background. Our investigation has led us to a remarkable degree of agreement with Leibniz, and in some respects we go further than he, for his designation of compossibility as the criterion which it is necessary to transcend is rather undeveloped, and our counterpart has far more structure.

The immediate importance of Leibniz for us was his view of physical space, as well as time, as relational. One can read the arcane writings and wonder how they link on to this basic position. The Ars Combinatoria seems to be a bit thin. True, it prefigured much of modern computing science, but Leibniz would have demanded much more than that. The answer must be that Leibniz was born in the wrong century, whereas Newton was born in the right one. Celestial mechanics played into Newton's hands: if we may imagine the facts of quantum theory having come to light first, things might have been very different. We can play with the idea of Leibniz being offered the relational space of relativity and the combinatorial aspect of the world exhibited by the quantum theory, and wonder just how far he would not have gone with such a marvellous springboard. Russell quotes Couturat[4] with great approval as saying "Leibniz's metaphysics rests solely upon the principles of his Logic, and proceeds entirely from them". He castigates Leibniz as being disingenuous for not coming into the open, as well as for adhering to the logic of the Schools. "Perhaps the most revolutionary conclusion in the whole book" (Couturat's) "is, that the principle of reason for all its trappings of teleology and Divine goodness, means no more than that, in every true proposition, the predicate is contained in the subject, i.e., that all truths are analytic." It is hard now, in the light of the criticism of Wittgenstein, to put oneself in Russell's position in which subjects and predicates and so on were completely commonsense and self-explanatory ideas, rather than having all the difficulties of metaphysical concepts. One can't imagine Leibniz falling into this trap of logicist absolutism. It is as though he knew that time was not ripe for his ars combinatoria because a metaphysics needs a subject matter which the physics of his day did not provide.

CHAPTER 3

Complementarity and All That

Mainstream attempts to present quantum theory as a coherent way of thinking all use the device of separating physical things into two classes: those which have to do with measurement or observation, and those which do not. The mathematics depends in an essential way on the device of treating these two classes differently. The device is fallacious because it is arbitrary whether we call a thing a measurement or not. The quantum thinker who struggled to the end of his life to understand and justify this device was Bohr, and therefore we consider we may bypass all the other writings on the mainstream foundations of the quantum theory by assessing the position of Bohr.

Bohr is credited with the remark that "truth and clarity are complementary", and Peierls, to whom the quotation is due, added that Bohr leaned heavily on the side of truth. Attempts at clarity about observation, in the sense of a brief and definite statement within the intellectual structure which we call quantum theory tend to run into the kind of difficulties which have occupied us at some length already. For example statements on the subject by Dirac are brief and admirably definite. However the more we strive for clarity, the more we find the difficulties thrown into sharp relief, and our 'clarity' is one which is obtained only at the price of being prepared to live with an underlying muddle. This particular sort of muddle, where there is a gap in the explanation which everyone tries to pretend they are happy with, has a special category in conventional theology, where it is called a 'Mystery': One eminent physicist took the line that one got into such muddles because one asked the wrong questions, and that if one did that it showed that one had not been 'properly instructed' — had not appreciated, in fact, what things were Mysteries.

Bohr differed from his contemporaries in the mainstream of quantum physics in not being prepared to temporize with an incomplete understanding of the basic quantum situation. However if our position of the last chapter is right, and the incompatibility of the classical view of measurement and any view which takes into account the way measurement is done in quantum physics is ultimate, then the endlessness of Bohr's search for a reconciliation was inevitable, and the accusation of unclarity, beside the point.

In this chapter we shall be concerned with the question: does Bohr's complementarity principle enable us to understand and therefore see as necessary the differences between the 'quantum object' (Bohr's phrase) and the object of classical physics? At the time Bohr wrote, these differences were usually seen to be expressed adequately by the uncertainty principle. For Bohr, much more explanation was needed. So far we agree with Bohr, but we shall conclude that even Bohr's profound critique did not issue in an explanation, and therefore that no completely satisfying understanding of the difference exists at present.

The doctrine called 'complementarity' is used to underpin the concept of observation in mainstream quantum theory. Presentation of it starts from the relationship of the particle picture and the wave picture, and goes on to a more technically articulated relationship which exists between certain pairs of dynamical variables. These appear as conjugate coordinates in the specification and solution of any single dynamical problem. The complementarity doctrine asserts that there is an exclusivity in the application of two such techniques or concepts at a given time, even though both are needed for the full understanding of the problem. Born[1] gives the following brief and authoritative account:

"The true philosophical import of the statistical interpretation ... consists in the recognition that the wave picture and the corpuscle picture are not mutually exclusive but are two complementary ways of considering the same process — a process whose accessibility to intuitive apprehension is never complete, but always subject to certain limitations given by the principle of uncertainty. ... The uncertainty relations which we have obtained simply by contrasting with one another the descriptions of a process in the language of waves and in that of corpuscles, may also be rigorously deduced from the formalism of quantum mechanics — as inexact inequalities, indeed: for instance between the coordinate Q and momentum P we have the relation

$$\delta Q \delta P \geq h/4\pi ,$$

if δQ and δP are defined as root squares ...."

In this account Born is more definite than most expositors in that he makes the Heisenberg uncertainty relation depend upon the more fundamental complementarity of the wave and particle pictures ('corpuscle picture', as Born calls it). However even he leaves ambiguous the question 'in which order does deduction proceed?' (Which is fundamental and which derivative.) He says that the complementarity principle leads to an uncertainty relation, and asserts that that is confirmed by rigorous quantum mechanics. Of course this would be fine if the more rigorous treatment included a more rigorous treatment of the wave/particle duality, but the treatment of that topic that appears in the above quotation is all the justification of it that he provides. Invariably other writers use uncertainty and the wave/particle duality in setting up quantum mechanics. Born's readers are left chasing round and round, and never sure at what point they are meant to break into the argument. Other writers are, as a rule, less clear than Born on this matter.

The evident — though perhaps never consciously expressed — invitation in such discussions as Born's is that one should build up support for the quantum-mechanical approach as a whole by deriving a little from each of an array of principles of which complementarity is one. Others are the Pauli exclusion principle, the commutation relations and of course the Heisenberg uncertainty. The student naturally wants to know whether these are independently necessary, or, if not, which entails which and in what order, but he gets no answer.

One might argue that this happens in classical mechanics. There, if we ask for a definition of mass, we are referred to statements which presuppose that we already know what force and acceleration mean; and vice versa; and so on round and round in circles. The closure of the system works, moreover. Everyone who is trained in physics knows exactly how to apply the classical theory and what constitutes a proper argument within it. The miraculous-seeming quality of the coherence is what we tried in Chapter II to draw attention to with our introduction of the term "theory-language". In the classical case we indeed have a closure of the definitions, which justifies us in starting from any of many equivalent points in our deductive treatment of any problem. Moreover, as happens with languages, every piece legitimately contributes to the meanfulness of the whole. The vital point however, in the case of the classical theory-language, is that principles which would be invoked in justifying any one piece would be consistent with those for all the rest, whereas in the case of the quantum theory, this consistency is just what is being called into question.

Bohr was not content to see complementarity as a sort of philosophical gloss on the wave-particle duality which might make the duality more acceptable to those who happened to like that sort of thing. He single-mindedly presented it as an autonomous principle separate from the more technical principles of the quantum theory, and requiring no justification backwards from the empirical success of the theory. On the contrary it was this principle which should carry the weight of the quantum-theoretical vision of the world.

The following short statement by Bohr himself appears in an essay entitled "Natural philosophy and human cultures".[2] "Information regarding the behaviour of an atomic object obtained under definite experimental conditions may, however, according to a terminology often used in atomic physics, be adequately characterized as complementary to any information about the same object obtained by some other experimental arrangement excluding the fulfillment of the first conditions. Although such kinds of information cannot be combined into a single picture by means of ordinary concepts, they represent indeed equally essential aspects of any knowledge of the object in question which can be obtained in this fashion."

This definition makes use of several principles which Bohr considered established in current theory, or which he considered he had himself established. Firstly, Bohr considers how information is obtained by an experimental arrangement. The units into which it was alone legitimate to analyze knowledge about the world of quantum objects are whole experimental procedures (the appropriate experimental arrangements being imagined separated from the rest of the physical surroundings of the experiment). This principle sounds arbitrary until we see it against the special operational circumstances of the quantum objects. It is part of what we mean by the term 'particle' in the classical way of thinking that there should automatically be the possibility of defining other particles in the neighbourhood of the first without making any special theoretical provision for their intrusion. If we could not assume this without question we should not be able to use the dynamical variables with their usual meaning. Now in the quantum domain this assumption is consistently and as a matter of fundamental principle invalid. If we wish to refer to a new particle then we must specify a new, and usually much more complex, theoretical background capable of describing the combined system. For Bohr, the right way to express this specifically quantum view was to stress the unity of observed entity and observing system, and indeed to insist that neither should be ascribed reality independently. Bohr's assimilation of the atomic object (to use Bohr's phrase) to the circumstances of its measurement is further emphasised by what he called the "quantum postulate". This, he says[3] "attributes to any atomic process an essential discontinuity or rather individuality, completely foreign to the classical theories, and symbolized by Planck's quantum of action". One sees from this quotation that Bohr saw the very discreteness or particularity of the quantum particle as something to be imagined quite differently from the way we imagine a classical particle. In particular, the automatic classical assumption about other particles just mentioned would presumably be an imaginative prop that Bohr would require us to renounce.

A second principle (II) which contributes to Bohr's idea of complementarity concerns the inevitability of the classical macroscopic description using the classical dynamical concepts, since — he argues — it is the only possible way of talking about the world in any of its aspects, including the quantum aspect. The second goes much farther than the first in the way it asserts that change from the classical language is for ever ruled out. Bohr was insistent on this strong prohibition. Any suggestion that we ought to be open to change at the basic level of the intuition of spatial events so as to get a new kind of description appropriate to the quantum object, seemed to him entirely fanciful. This position of Bohr's seems at first sight like that of the naive realist who does not question the prevalence of classical language because it has not occurred to him to do so. In fact the positions are poles apart. In his positivistic attitude to the language of physics Bohr was stressing the all-embracing character of what we have called the 'classical theory-language', and by no means saying that the dominance of classical description was an expression of common sense.

A good deal has been written about the influence of idealist ways of thinking that may have made Bohr feel he was on the right track in insisting on the classical language as though it were a necessary form of thought, or at least a precondition for all thinking that could be labelled 'physics'. In particular Bohr may have seen an analogy between the part played by the classical language and the synthetic a priori place of space and time in the Kantian philosophy. Again the complementarity idea certainly recalls the antinomies of Kant and the inevitable opposition of pairs of concepts in the dialectic of later German idealist philosophers. See Petersen[4] for discussion of these and related points.

The last component (III) that we always find in Bohr's statements of the complementarity principle is that of incompatibility. We already have the unity of the operations and the language that go to make up a measurement: we have the restriction on the scope of that language to that which is current in the classical understanding of the world: now we are to understand that there will typically be more than one such description required to present the essentials of any given quantal situation, and that these will consistently appear so that the provision of one will consistently prevent the provision of the rest. As Petersen puts it: ".... the experimental arrangements that define elementary physical concepts are the same in quantum as in classical physics. For example, in both cases the concept of position refers to a coordinate system of rigid rulers and the momentum concept refers to a system of freely moving test-bodies. In classical physics these instruments can be used jointly to provide information about the object. In the quantum domain, however, the two types of instruments are mutually exclusive; one may use either a position instrument or a momentum instrument, but one cannot use both instruments together to study the object."[4]

WHY NOT? It is very difficult even to imagine what it would be like to argue in favour of Petersen's assertion, let alone actually to produce the argument. What sort of thing could it be that would prevent one kind of instrument being used because of the presence and use of the other? Would it be like crossing the critical-mass boundary in neutron emission so that bringing up the second instrument would cause an explosion? Or would there be consistently destructive interference? Or what? Again, would the argument be that it was the successful operation of the one instrument that must inhibit the operation of the other? And then, what would the mechanism of the interaction between the two be? It is obvious that if one restricts oneself to classical arguments then there is no reason why one should not, for example, construct measuring techniques which measure momentum and position and other dynamical variables as well in indefinitely complex relationship. Indeed it is notorious that, far from its being the case that simple dynamical variables force themselves on the attention of the experimenter, his ingenuity is always stretched by the need to provide experimental techniques that exhibit those conceptually simple properties of a system that theory demands.

Of course, we shall be accused of being perverse in making these remarks: of deliberately missing the point. The proper way to understand Bohr, his apologist will say, is to see that the system behaves as though there were this incompatibility. But if he says this he is retreating from his claim that the complementarity idea was an explanation of the characteristic quantum situation. For it to be an explanation and not merely a gloss, we have to see the incompatibility in action and understand where the exclusivity comes from.

There is one argument which might be put forward to show that a form of complementarity is a necessary consequence of the conditions of measurement. It is much cruder than any advanced by Bohr, and yet it has a sort of logic and ought to be mentioned if we are to claim that we have exhausted all arguments that support the current epistemological doctrine about observation. The argument runs: "Suppose that we measure position. Then if we measure momentum this must be by observing scattering of some sort of 'test-particle'. However we don't have test-particles which are small compared with the particle being observed (as we always do classically). Hence the momentum measurement must disturb the position, and this is merely a fact about measurement." One has to admit that a sort of uncertainty and a sort of limitation on simultaneous measurement do follow from the assumption of identical particles of finite size as the smallest available test-particle. The objection to the argument is that it is circular. The finiteness of test-, and all other, particles is supposed to be a consequence of whatever theory we propose to understand the quantum world, and it is therefore not available to use in the argument to establish that theory. The current quantum theory is actually open to the same objection because of its arbitrary imposition of finite constants — notably h itself — though the crude realism about particles is covered up a bit.

The conclusion we reach is that Bohr was not successful in using complementarity as the deductive basis for the quantum theory. The most we can say is that he may have showed that the complementarity philosophy was needed to make the wave-particle picture, and derivatively the uncertainty relation, comprehensible assumptions by exhibiting them as examples of something which transcends physics.

At a recent conference[5] speakers looked at different points in current science at which the hitherto absolute categories of 'nature' and 'cognition' were becoming blurred. There is no doubt that Bohr's point of view does involve this blurring, and a main aim of the conference was to see if the idea of complementarity would provide guidance in this new and unfamiliar terrain. The conference provided an opportunity for a contemporary reappraisal of Bohr's position. Unfortunately from our point of view, none of the expositions of the complementarity principle from the point of view of technical physics claimed to explain the incompatibility of physical mechanisms in observation (which is what we have argued was needed to fulfill Bohr's aspirations).

Let us end with something more positive. Earlier, we tried to explain Bohr's stress on the unity of the whole experimental procedure that leads to knowledge about a quantum object by bringing in ideas which are naturally to the fore in high energy physics. A quantum measurement does not presuppose the potential existence of a background of related experimental results in the way that a classical measurement does. Even for such a (classically) simple case as the measurement of two momenta at contiguous points at high energies, we require a quite different experimental arrangement from what is needed for the single measurement, and one that is usually of a much greater order of complexity. To see that this difference is a matter of principle and not merely of practicality is to come near to saying, as Bohr does, that different ways of measuring imply that different things are being measured. Indeed it would be only a small step to build the whole quantum-theoretical concept of measurement round the application of this idea to momentum and spatial position. Then one would have the uncertainty principle, in all essentials.

This way of looking at quantum measurement might well have been sufficiently radical for Bohr (who always stressed that his quantum postulate was something quite new on the horizon of physics) but it would not have satisfied him because it cannot be made to follow from the classical concept of observation. One might even say that it was Bohr who led the way toward recognition that the quantum theoretical view of space and of the notion of locality must be relational rather than absolute, because of the impossibility of separating the atomic phenomenon from the special circumstances of its measurement including an account of the whole experimental arrangement.[6] However we should also have to say that Bohr had definitively blocked off what we have seen as the logical way forward for a recognition of the relational nature of observation by his insistence on the classical intuition and the classical description as the only starting point. In what follows in this book we shall be regarding measurement from a specifically quantum point of view, and regarding the classical case as less basic; and so at this point we part company with Bohr.

CHAPTER 4

The Simple Case for a Combinatorial Physics

The quantum theory had its origins in the pressing need to give some theoretical account of discrete aspects of the world which experimental knowledge was making it more and more impossible to ignore. It is now about a century since this innovatory episode, and the intellectual capital that is locked up in the total enterprise that has quantum physics as its centre is vast. Yet it is not impertinent to ask how far the explanatory effort to understand the discrete, as it was first conceived, has succeeded.

The critical step of the early days of the theory was Planck's imposition of a mathematical restriction to the range of allowed values of the energies of oscillators in the radiation field. The story of the scientist who, on hearing the theory expounded at an early meeting, left, with the remark "Gentlemen, this simply isn't physics" has become part of the 'memorable' history of physical science. At that early stage there would not have been serious disagreement about the ad hoc character of Planck's innovation: people accepted that a highly successful empirical formula had been found; took it for granted that no reason had been given for the success; and waited for further understanding of the situation to provide a 'proper' explanation. Formally, the hypothesis of discrete energy levels was simply inconsistent with continuum physics, yet it seemed to work. However, through exhaustive discussion of the dynamical possibilities allowed by classical mechanics in the interaction of radiation with matter in the two decades which followed (and compellingly summarized in Jeans' "Report on radiation and the quantum theory") it finally had to be accepted that no theory of a classical sort could accomodate the hypothesis of the discrete levels. The more reactionary of the commentators on Planck's hypothesis seemed to have been the nearer the mark.

As the quantum theory developed, the attitude of the majority to the existence of discreteness within a theoretical structure changed. Most accepted that the modern (1926–30) form of the theory was an intellectual structure within which discrete and continuous quantities could cohere. However arguments to be found in the literature and summarized in earlier chapters have made this point of view seem, at best, complacent: we have been able to find no unifying framework of thinking. Mainstream attempts at such a framework invoke the measurement process, but no one was able to make sense of that. Fundamental enquiry on this matter came, almost as though by common consent, to be left unresolved from the time of the controversy between Bohr and Einstein. Rather than follow the subsequent history of that controversy into the currently widespread concern with non-locality and Bell's theorem, we shall return to the unresolved spirit of that controversy, with a root and branch attack.

The quantum theory is about discrete processes, but has a history of being presented against a background of continuum mechanics. The two are actually incompatible, and if they are treated as though they have by divine decree to be made to look compatible, endlessly proliferating complexity will be the price that has to be paid. The chosen battle-ground for the conflict of ideas between Einstein and Bohr was the Einstein–Podolsky–Rosen Gedankenexperiment, but this particular battle was only meant to give precise point to what Einstein thought was a general departure from scientific method. His demand was for what he called 'completeness' in scientific description. The matter is put by Aharanov and Petersen[1] like this: "The mathematical formalism of quantum mechanics imposes restrictions on the definability of physical variables in a given state of a system. The relation of these restrictions to the possibilities of measuring the variables has been a main issue in the discussion of the foundations of the theory." Einstein's position was that no such restrictions were admissible.

Aharanov and Petersen continue: "Heisenberg [W. Heisenberg, *The Physical Principles of the Quantum Theory*, (Chicago, 1930). See also Heisenberg's article in S. Rosenthal (ed.) *Niels Bohr* (Amsterdam, 1967) 92] and Bohr [N. Bohr, *Atomic Theory and the Description of Nature* (Cambridge, 1961)] claimed that there is complete agreement between the possibilities of definition and the possibilities of measurement. Against this view, Einstein [See the discussions between Einstein and Bohr in P. A. Schilpp (ed.), *Albert Einstein, Philosopher-scientist* (Evanston, 1949) 199] repeatedly tried to show that measuring procedures are available which permit us to specify physical variables with greater accuracy than that allowed by the quantum formalism. In contrast, Landau and Peierls [L. Landau and R. Peierls, *Z. Phys.* **69** (1931) 56] argued that in quantum field theory the methods of measurement are far more restrictive than the limitations imposed by the formalism. However, in each case [N. Bohr and L. Rosenfeld, *Dan. Vid. Selsk. Math. fys. Medd.* **12**, (1933) 8, and Schilpp, (ibid)] a closer analysis of the measurement problem restored the equivalence between definability and measurement."

One does not need to be strongly partisan on Einstein's side to ask whence comes this miraculously precise identification of description and measurement? What strange pre-ordained harmony assured it? The answer is not far to seek: the quantum theorists always insisted that the quantum theory contains measurement as part of its structure, and so it is not so surprising that they were able to restrict description to that which measurement, as defined in the theory, allowed. To put it differently, 'description' had become 'quantum description', and measurement went with it. In their hands however, measurement had become something quite different from the classical procedure which stood outside theory as its arbiter. This it was that Einstein felt to be unscientific. However he seems to have slipped unwittingly into a discussion which used the restricted, quantum theoretical, kind of measurement. The Einstein–Podolsky–Rosen paradox began this trend, and the 'clarification' offered by later forms of the paradox, like that of Bohm, could be said to have clinched the circular interdependence of description and measurement. Bohr and his associates are usually considered to have won the argument, because wherever there has been a clash between expectations based on classical ideas about the spatio-temporal framework and the detailed consequences of quantum theory, the latter has always been vindicated.

However there is an important sense in which Einstein was right to have taken the stand he did, and this is why it is worth going back so far into history. The quantum theorists had changed the concept of measurement most fundamentally, but were continuing to expect the advantages of old and new to operate together. Probably some sort of deep-seated feeling about the unity of the physical world underlay their failure in critical acumen. Einstein himself clearly thought that one view of measurement or the other had to be the truth, and so it was that the opportunity of that moment, to use the incompatibility of these two important concepts to find deeper foundations, was lost.

The pre-quantum view of measurement was inseparable from our laboratory-scale experience of positions and movements of objects in space and the changes in those, as time passes. The close mutual dependence of these concepts reached its apotheosis in the theories of relativity, and therefore it was natural to express the novel character of quantum events through their contradiction with the metric connections of relativity. However clarity earlier on would have led one to expect such a result instead of seeing it as paradoxical. Similarly people would not have found it so strange that unification of the quantum theory with relativity continues to elude us.

If measurement in the quantum world is no longer based on macroscopic spatial experience, what is it based on? The answer is: it is based on counting.

It is our aim in this book to present a form of quantum physics in which all relationships are combinatorial. The theory provides some combinatorial relationships which give the results of measurements correctly in terms of counts of experimental occurrences, and which therefore serve as a test of the correctness of the theory. And these theoretically calculable counts are special cases of wider ranges of counts which just describe how the world happens to be, and which no one would expect to be able to calculate. Conventional measurement is regarded the same way.

The possibility of a change from conventional measurement to counting is an example of a change from one scale of measurement to another, and there are three scales of which counting is the last. This classification of scales of measurement seems first to have been set out by N. R. Campbell.[2] The scales are best thought of as a sequence of increasing absoluteness. All of the scales assign magnitude to one physical attribute. The first scale does no more than set measurements in order, and may be called the 'ordinal scale'. Some process of comparison is agreed upon which can be applied to every pair of the things to which magnitude is to be assigned in respect of the attribute under discussion. By using this process we select a set of standard cases, of each pair of which we are able to say either that the first is greater than the second, or that the second is greater than the first. These standard cases form an ordered set, and every other case can be placed between two of them unambiguously by using the same comparison process. In this way the whole field is classified into an ordered set of equivalence classes by the ordinal measurement process. It will easily be understood that for the scale to be useful in practice, adjacent standard cases should be in some intuitive sense discriminable, but only just: it would be inconvenient if several other standard cases could be interpolated between any two of them.

Examples of ordinal measurement scales in common use are Moh's scale of hardness, and the Beaufort wind scale. In the former, the comparison process consists in scratching one of the members on the scale with another. A set of standard cases is selected starting with the hardest known substance, and these cases are given the ordinal numbers from 1 to 10, so that a given substance will always scratch a substance of greater ordinal.

There is no unit in the ordinal scale. Units appear with the 'ratio scale'. Ratio scales become appropriate when one can give some sense to the idea of different points on the scale being equal distances apart. Temperature is a historically important example of a ratio scale. Over a wide range of temperatures change in choice of working substance does not grossly upset the consistency of the measurements. The unit in the ratio scale is arbitrary — being a fraction, such as one hundredth, of the change of the working substance between two easily reproduced fixed points such as the freezing and the boiling points of water.

Almost all the results of experiments in the more traditional physics are expressed in ratio scales, and there is a close connection between measurement on a ratio scale and the idea of a continuous mathematical variable. The connection comes from the arbitrary character of the choice of unit. Since it would be quite wrong to attribute any theoretical significance to the steps marked off by the unit, it is necessary to find a way of giving equal significance to the intervening points. This is done without appeal to any new principles by using the formation of ratios of units to measures on one scale, which is already in use, a second time. Now we form ratios of different measures in terms of the same unit. This gives us the rational numbers to represent measurement, and is as far towards the continuum as we need, or are able, to go. It is true that in conventional continuum physics one extends the rationals to the reals, but this is only for mathematical convenience.

We can take the construction of measurement scales one last stage in progressively reducing arbitrariness by getting rid of the arbitrary unit. We call the result the *absolute scale.* And the absolute scale identifies measurement with counting. The logical possibility of taking this last step is very important for our present argument because in taking it we answer the question with which we began this chapter. The discrete quantities whose discovery initiated the quantum theory are explained if measurement is equivalent to counting. From this very simple and classical position of analysing the quantum measurement problem in terms of scales of measurement, it is natural to argue that all the difficulties arise because we have a deep-rooted preconception that all physical measurement is on a ratio scale whereas really (as we now postulate) that part of it which is characteristic of the quantum world should be on an absolute scale.

The most important parts of quantum physics — particularly those concerned with particle scattering experiments — have moved a long way towards giving counting (of particle processes) pride of place as the paradigm of measuring technique. This change supports our position on absolute measurement. However we shall use much older ideas for the moment. The immediate problem is this: if measurement is on the absolute scale, how does the fact show itself? It shows itself in the quite new and non-classical type of discreteness of the quantum attributes. Everyone would agree that the discreteness of quantum attributes is systematically and most strikingly unlike, say, the discreteness of the mass of cricket balls all of them five and a half ounces give or take a bit.

There are numerical magnitudes in modern physics that seem tailor-made to be the stepping stones into the absolute scale. These are the ratios of atomically or cosmologically important constants whose ratios are independent of the choice of the arbitrary, or ratio-scale, units. We call them dimensionless constants: they are in fact pure numbers. These pure numbers include such things as the ratios of the masses of the elementary particles. Other cases will incorporate both atomic and cosmological constants. It is convenient to call these latter the *scale-constants* because they can be thought of as setting the relative scale of physical phenomena.

If the scale-constants are looked at from the point of view of a ratio scale of measurement then one will regard the natural fundamental constants which are their factors as the theoretically primary quantities, with the scale-constants as derivative. From the point of view of the absolute scale, things are the other way round, and the scale-constants are primary. The practical consequences of this change of viewpoint are profound. The former view goes with traditional thinking where one is accustomed to each natural dimensional constant playing its part as a parameter in some major division of physics separately from the others. (The electronic charge, for example, in electromagnetic theory; Planck's constant in another experimental context, and so on.) Thus we are used to thinking of the scale-constants as the results of bringing them together into dimensionless ratios whose values just happen to be what they are. In sharp contrast, if the scale-constants are primary, then we have to imagine that the different branches of physics in which the factors of the scale-constants play their respective parts have developed into the form that they have, with the fundamental dimensional constants playing the role they do, to fit in with the overriding constraint defined by the scale constant. This, on the face of it, is a startling change: it will be worked out in detail in later chapters.

The idea that there may be some reason lying at a more profound level than existing physical theory which prescribes the values of the scale-constants, was advanced in particular by Eddington[3] in his writings around the period of World War II. We can justify Eddington's position by the following argument, though he does not seem to have put it like this himself: (a) the scale-constants have numerical values which are determined in some way, prior to the determination of their factors — the 'constants of nature' — so that it is the former which impose constraints upon the experimental values of the latter, and (b) the reasons why the scale-constants have the values they do are, in principle, discoverable. We shall refer to this statement as 'Eddington's conjecture'.

Eddington did try to provide reasons why the scale-constants should have approximately the values they are observed to have. He did, in effect, assume that the calculations would be combinatorial. His attempt found its combinatorics in the structure of space-time, and, derivatively, the numbers of components of quantities analogous to the energy-momentum tensor under different constraints which reduced the number of components. The bare bones of his argument can be seen by writing the tensor-like quantity as a square array which is the product of two vectors with a $3 + 1$ structure. There are 16 products of the elements pairwise. Six of them mix time-like and non time-like elements, and ten do not. Repeating the process we can form 256 products of the elements of the array pairwise. Some of these mix time-like and non-time-like elements, and some do not. There are $10 \times 10 + 6 \times 6 = 136$ which do not. Eddington saw this number as the origin of the fine-structure constant, which is the reciprocal of the scale-constant associated with the electromagnetic field — $2\pi e^2/hc$. The experimental value of this constant is much nearer to the reciprocal of the integer 137 than to 136, and Eddington offered reasons for adding 1. These carried little conviction.

s	s	s	t
s	s	s	t
s	s	s	t
t	t	t	s

This calculation is quite different from the one we give later: indeed if Eddington's calculation is correct then ours is wrong simply for the reason that you can never maintain two theoretical accounts of some experimental finding which have different and irreconcilable premises and arguments. The way the matter appeared to us was that the case for Eddington's conjecture was so strong that one was bound to evaluate Eddington's calculation, and, if it could not be supported, replace it with something which could. We therefore set out to find conditions under which dimensional structure could consistently be used combinatorially to achieve Eddington's purpose. We came to the conclusion that Eddington's argument was flawed because the only existing definition of physical dimensionality — which in the absence of any contrary statement by him he must be presumed to have been thinking of — was based upon a ratio scale of measurement, and Eddington had not undertaken the task of providing one suitable to be used combinatorially.

The Eddington calculation was rejected by physicists generally, though for different reasons. The fact that the argument was sketchy had less influence than the view that Eddington had somehow departed from proper scientific method. The remark that was heard on all sides was that the theory was "a priori" and therefore an affront to scientific integrity. Numbers were being assigned to things in the world because of some considerations that arose in the mind, instead of through the results of experiments. This charge is, on the face of it, odd. What, one may reasonably ask, is the aim of scientific theorizing if it is not to find a deductive basis for numerical relationships that have hitherto been known only from experiment?

Moreover — and now we put the case for a combinatorial physics in its simplest form of all — once the new kind of non-classical discreteness of the quantum is appreciated, it must have been clear to everyone that some numbers had got to be injected into physics somehow and somewhere. This conclusion, at any rate, has become progressively more clear as the quantum theory has developed to its present form. Eddington was only anticipating the present generation of high energy physicists in trying to get the underlying combinatorial structures out of space-time structure. All the current particle theories try to get their symmetry structures ultimately from groups many of which have a space-time interpretation (Lorentz, Poincaré, and the like). Oddly enough, the trail which was started for us by Eddington's work in this direction has finished up with a new combinatorics which does not derive at all from space-time structure, and has to deduce the latter from something more fundamental.

One concluding point: We have argued for the general position of which Eddington was the first advocate, that there must be combinatorial structures of some kind at the heart of physics, which provide experimentally verifiable numbers. We have contended that this position becomes reasonable if measurement can be reduced to counting. However the numbers (the scale-constants) are not directly the results of counting anything, and so there is a gap in the argument to be filled. This necessary completion of the argument has to be left to later, when an intimate connection between the scale constants and counts of particle scattering processes will be developed. The fact that the experimental values of the scale-constants are not exactly integral will also be dealt with in detail.

CHAPTER 5

A Hierarchical Model —
Some Introductory Arguments

A particular mathematical model, whose most important feature is due to A. F. Parker-Rhodes,[1] is central to the theory we shall henceforth be putting forward. In this chapter we give several presentations of the same model from different points of view which have seemed the best way to approach it to different people over the twenty-five years it has been around. In the next chapter we provide a rigorous treatment. We begin with the first attempt to formulate these ideas.

Argument 1. Similarity of position

In this argument, which is the oldest we present, dating from 1954, a property of elements of a certain formal system is used to express a kind of symmetry which the authors[2] thought at that time to be an essential feature of dimension structure. It became clear later that the same logic was important in explaining how elementary particles, seen as bundles of attributes (quantum numbers) have the property of being able to be substituted in different situations and yet be recognizably the same.

We consider a system with a single binary operation, denoted by concatenation, which is such that any equation in the system is converted into another (true) equation in the system by every permutation of the elements. This property is called similarity of position. Then, if $ab = c$, it must also be the case that $ba = c$, so that the binary operation is commutative. If $a(bc) = d$, then $c(ab) = d$, so $a(bc) = c(ab) = (ab)c$ using the commutativity; thus the operation is associative. Again, since $ab = ba$, it follows that $aa = bb$, so that all elements have the same square, e, say. And since $aa = e$, $ae = a$, and therefore e is an identity element. We evidently have an Abelian group in which every element is of order 2 and, by a well-known result in group theory, the system is

$$S = C_2 \times C_2 \times C_2 \ldots ,$$

a direct product of n copies of the cyclic group of order 2. The system must therefore consist of e with $2^n - 1$ other elements.

This argument captures the desired property of substitutability, but is mysterious in other ways. There is, in particular, no physical explanation of why the system should have a single binary operation. However it is useful to have the argument in mind because it is isomorphic to another which comes by an apparently different route. The isomorphism is set up by noticing that the binary operation is commutative (as well as associative) and so might more naturally be written as addition with e written as 0, and $a + a = 0$ for all a. In fact, the elements of each copy of C_2 could be written [0,1], the group operation being addition mod 2, and the corresponding elements of S being then strings (vectors) over Z_2. This structure will be noticed again in argument 2 and elsewhere.

After the authors had discovered argument 1, progress was slow until A. F. Parker-Rhodes[1] produced an apparently different argument. We present this next in a treatment which is as mathematically compressed as it can be, and which is near the form in which it was discovered — hindsight being used to avoid confusion in the original work.

Argument 2. The original combinatorial hierarchy

(The reader concerned mainly with the physics may read paragraphs 1,2 below, and then find the mathematics concluded in its shortest form in paragraph 3.)

1. The algebra starts life by representing an imagined idealised physical process which asks of any two things whether they are the same or different. Consistently with ordinary English usage we call this *discrimination*. The process is primary. We have to say we start with just two things, a, b, but anything else we could know about them other than their distinction would have to be got from further processes of the kind we are formalizing. Writing S for 'same' and D for 'different', the possibilities of change can be written

A Hierarchical Model

	a	b
a	S	D
b	D	S

One can code (a,b) as $(0,1)$. That is a mere matter of notation. Also one can code (S,D) as $(0,1)$ or else (S,D) as $(1,0)$. The choice makes no essential difference but it makes better sense if "same" = "no difference" is written 0. We now have a set of 2 symbols closed under a binary operation, written "+".

	0	1
0	0	1
1	1	0

This is discrimination.

2. We now have a problem. '0' is special, because it shows the "sameness" of things when it appears in the answer. We need to put it on one side, reserving it for that purpose, and that leaves only '1' as the code for objects, and since we must be able to look at at least two objects we need another symbol, say '2'. One can write these as binary strings, in order to stress that everything is an existence symbol or its opposite, and it is easiest to do so in reverse notation, so that 1=(1,0), 2=(0,1). Then you are in the same system of 0,1 as in argument 1, and you take

$$(p,q) + (r,s) = (p+r, q+s)$$

so defining discrimination between strings. It is discrimination because $(p,q) + (r,s) = (0,0)$ if and only if (p,q) and (r,s) are the same. Hence $(1,0) + (0,1) = (1,1)$. Because we are working with strings it is useful to have a different notation in which we note down the different places with a 1. Thus $(1,0,1,1)$ is written $(1+3+4)$.

Now when we discriminate: $(1) + (2) = (1+2)$, or

$$\begin{bmatrix} 1 \\ 0 \end{bmatrix} + \begin{bmatrix} 0 \\ 1 \end{bmatrix} = \begin{bmatrix} 1 \\ 1 \end{bmatrix}.$$

The symmetry of '+' (computer people call it reversibility) means that $(1)+(1+2) = 2$ and so on. Discriminating any pair of different strings gives another in the set (of 3 in this case). Such a set is called a *discriminately closed subset*, or dcs (it is a "sub"set of the set of all binary strings). This idea of closure under the fundamental operation of discrimination is obviously a natural one, but dcs's (dcss) are important for other reasons as will soon appear. Starting with two things $\{1, 2\}$ how many dcss are generated? We recall that the definition of dcs is: $a = b$ is in the set for any two a, b in it, so single elements qualify by default. There are then 3 dcss: $\{1\}, \{2\}, \{1, 2, (1 + 2)\}$.

3. In the sixties the binary string notation with the 'existence symbols' 0, 1 was used by Parker-Rhodes to formalize ideas due to the authors of a hierarchy of levels in which operations upon strings at one level became the elements of a next level, and so on. Subsequent work developed from this formalization, and we present it here as briefly as possible so that it is short enough for the non-mathematician to hold it in his mind. It contains the essential mathematical trick which enables the important numbers to be calculated, even though the fuller argument (points 1, 2 etc., and going on to Chapter 6) is needed to get rid of the sense of arbitrariness.

There are 3 dcss in the set of 2-strings. Consider non-singular linear operators which preserve these in the sense that each string of the subset is transformed into itself and no other string is transformed into itself. We can say that these operators preserve discrimination. Three 2×2 matrices with 0 or 1 in each place can be found to perform these operations, one for each set, and the set of three is unique. Treat these three matrices as elements of a new level, and to emphasize this change of conception write them (in any consistent way) as 4-strings. The whole level is now formed by constructing all the 4-strings possible using the discrimination operation between different elements, so that the original three form a basis. There are $2^3 - 1 = 7$ such strings in all. These seven strings are organized into 7 dcss bringing the total number of dcss to 10. This construction of levels can be continued by finding 7 non-singular 4×4 matrices; one for each dcs. At the next level, when these matrices are seen as 16-vectors there will be $2^7 - 1 = 127$ strings. The set of seven operators is no longer uniquely determined. In fact there are 74088 such sets. There is one difference here from the earlier level change. The strings will usually be organised into 127 dcss, making 137 in all, but for a minority of the 74088 cases (about 10,000) a smaller number of dcss will be produced, because, in the algebraicists' parlance, the 7 operators are not linearly independent. Parker-Rhodes gave the construction in the form "Choose the 7 operators so that the 127 dcss are produced at the next level".

Continuing the construction, it is not obvious that there are 127 linearly independent matrices (out of a maximum possible of $16 \times 16 = 256$) which perform the operations required by this algorithm, but it has been shown that there are. At the next level things are different. The number of operators required is $2^{127} - 1$, which is of the order of 10^{39}, and there is no hope of finding these (linearly independent) out of $256 \times 256 = 65536$. Hence the construction stops.

The authors and Parker-Rhodes intuited that the construction was of sufficient generality (though this took a long time to show fully) to justify our attaching physical significance to the sequence of cumulative sums: 3, 10, 137, order 10^{39} (with its remarkable cut-off). This is because of their similarity to the sequence of dimensionless numbers on which physical magnitudes are based and which are called the coupling constants of the basic fields.

4. We now return to our more careful development. The sets of strings (there will soon be more strings and longer ones) are functioning as labels for things. If there are two things it is immaterial whether one is 1 and the other 0 or vice-versa. It makes sense to look at permutations of the labels, or equivalently of the elements of the dcss. Here is a table for the dcss of the 6 permutations. (HEALTH WARNING: this dcs is not typical of all larger ones.)

	1	2	1+2	operator	inverts
I	1	2	1+2	[1, 2]	←
II	1	1+2	2	[1, 1+2]	←
III	2	1	1+2	[2, 1]	←
IV	2	1+2	1	[2, 1+2]	No
V	1+2	1	2	[1+2, 1]	No
VI	1+2	2	1	[1+2, 2]	←

You can specify a permutation by saying what 1 and 2 go into. Then $1+2$ has to go into the remaining element.

5. Now, following up the health warning, the fact that there is only one remaining element to fill in, and that we are dealing with a dcs, means that all the permutations automatically preserve discrimination. That means, if elements a, b are permuted by one of the permutations p, say, to $p(a), p(b)$ then it will also be true that $a + b$ is permuted to $p(a + b) = p(a) + p(b)$. This justifies the use of linear algebra in the Parker-Rhodes algorithm (in the separated part above). If you think of addition instead of discrimination, then p is (automatically) linear, and so if $1, 2, 1+2$ are thought of as column vectors then the permutations are effected by non-singular square matrices. The "operator" column in the table ties in with this: what is put into the square bracket is simply the list of what 1, 2 go into. This then corresponds to a square matrix. The notation is simply to list the columns from left to right so that $[2, 1+2]$ means

$$\begin{bmatrix} 0 & 1 \\ 1 & 1 \end{bmatrix}$$

Calculations using the matrix algebra have advantages and disadvantages. For example in calculating Av where A is a square matrix $A = [a, b]$ say, and v is a vector (i.e. is 1, 2, or $(1+2)$,) the notation suggests more than is really happening, for v is simply selecting a column of A. If v is 1 it is the first; if 2, it is the second, and if $(1+2)$, it is their discrimination. So

$$[2, 1+2]1 = 2, \qquad [2, 1+2]2 = (1+2), \qquad [2, 1+2](1+2) = 1 \ .$$

On the other hand the matrix notation exhibits the way a number of degrees of freedom arises from the permutations: the matrix has $2^2 = 4$ elements.

6. Parker-Rhodes noticed that of the 6 perms listed, two are "unfixers", as Amson put it. IV and V change every element, three leave one element unchanged, and one leaves all three unchanged. As a consequence, to each of the three dcss mentioned there is a perm which corresponds to it in the following sense: that the perm leaves each element of the dcs and only those unchanged. (NOTE: three perms and not four because IV, or $[2,1]$, leaves unchanged $\{(1+2)\}$ which is not one of the dcss under consideration.) Thus you could say that the perm characterises the dcs. We notice that it will not at all do to try to characterize the dcs by a perm which leaves it, as a set, unchanged whilst possibly permuting its elements among themselves. For example the largest dcs $\{1, 2, (1+2)\}$ is left unchanged by all perms.

This is the most important aspect of dcss — they can be described either by their elements, as a set, or by the single perm which characterizes them. The next step is even more important. If p, q are any two perms there is a natural discrimination between them giving an operator (not necessarily a perm) $(p+q)$, defined by giving the operation of $(p+q)$ on any vector v as

$$(p+q)v = p(v) + q(v) \ .$$

This is a discrimination since if and only if $p = q$, $(p+q)v$ will be 0 for all v. In other words the perms are generators of a discrimination system, as we said in the short treatment "at the next level". The first stage of the construction is now over and finishes with the production of three perms.

7. At the next level we begin with 3 elements. They are perms or equivalently square matrices, and we want to repeat the process. In the short treatment, following Parker-Rhodes, you convert the matrices which are "bit-arrays" to "bit-strings" by any convenient rule, for example by standing columns on top of one another. Thus:

I:
$$[1,2] = \begin{bmatrix} 1 & 0 \\ 0 & 1 \end{bmatrix} \to \begin{bmatrix} 1 \\ 0 \\ 0 \\ 1 \end{bmatrix} = (1+4) \text{ or } 1$$

II:
$$[1,(1+2)] = \begin{bmatrix} 1 & 1 \\ 0 & 1 \end{bmatrix} \to \begin{bmatrix} 1 \\ 0 \\ 1 \\ 1 \end{bmatrix} = (1+3+4) \text{ or } 2$$

III:
$$[(1+2),2] = \begin{bmatrix} 1 & 0 \\ 1 & 1 \end{bmatrix} \to \begin{bmatrix} 1 \\ 1 \\ 0 \\ 1 \end{bmatrix} = (1+2+4) \text{ or } 3$$

In fact the arbitrary character of Parker-Rhodes' rule shows that it does not have to be done: you could simply label the perms with fresh labels.

8. How many dcss do these generate? The greatest possible number will be $2^3 - 1 = 7$. You calculate this by asking for each dcs whether 1 is in it or not. The answers to these 3 questions then determine the other elements. Note that $2^3 = 8$, and three no's is not allowed since something must be in. Is it actually 7? Yes; by doing it explicitly in matrix notation. (This will be different at the next stage.) For each of the seven ask for a perm which (a) characterizes it, and (b) leaves discrimination unchanged. We observe that (b) was satisfied automatically at the first level, but not now. An example shows this. Take the dcs $\{1, 2, (1+2)\}$ out of $\{1, 2, (1+2), 3, (1+3), (2+3), (1+2+3)\}$. As far as the perms that characterize this set and do leave + unchanged are concerned it suffices to give the value they turn 3 into. This cannot be $1, 2, (1+2)$ or it would not be a perm, and it cannot be 3 since it is to characterize a dcs not containing 3, so there are $7 - 4 = 3$ possibilities, which we list:

	1	2	(1+2)	3	(1+3)	(2+3)	(1+2+3)
I	1	2	(1+2)	(1+3)	3	(1+2+3)	(2+3)
II	1	2	(1+2)	(2+3)	(1+2+3)	3	(1+3)
III	1	2	(1+2)	(1+2+3)	(2+3)	(1+3)	3

These are not all the perms because there are 4 elements outside the set and the total number of perms is simply the number of unfixing perms on 4 members, which is 9, so there are $9 - 3 = 6$ perms which do not preserve +.

9. The table above shows that there is no longer a one-to-one correspondence betwee dcss and characterising perms. Here, for example, there are 3 perms all characterizing $\{1, 2, (1+2)\}$. For the 1-element dcss there are more. (In fact 14 preserving discrimination as well as others which do not.) Therefore the next step was put by Parker-Rhodes in the form of a choice: choose one perm to characterize each of the 7 dcss. These perms are the elements for the next level. We have now chosen in all $3 + 7 = 10$ perms.

10. The greatest possible number of dcss at the next level will be $2^7 - 1 = 127$. Since the perms at the next level down were, in the matrix version, 4×4-matrices, they come up at the higher level as 16-vectors. The 127 dcss, if they exist, are sets of 16-vectors. Will there be 127 or fewer? It depends on the 7 16-vectors generating. If whenever you discriminate any number of different ones you never get 0, you will acheive 127 elements in all, but if 0's come up, you will get smaller sets. Parker-Rhodes put it this way: choose the perms so that they generate 127 dcss. We would say now: allow all possibilities indifferently; some of them will generate 127 dcss.

11. Now things are just recursive, but it is at the next level that blood was sweated in the past. Can one choose 127 perms to characterize the 127 dcss (these perms will be 16×16 matrices, in the matrix notation) so that they, in turn, generate $2^{127} - 1$ or approximately 10^{38} dcss? If so then one will have constructed 137 perms. The proof was difficult: one has to show, in matrix terms, the existence of 127 linearly independent non-singular matrices in a space of matrices of dimension 256 which satisfy the characterization requirement. Since that time, a combinatorial discussion has made it easier and shown that in fact dimension $12^2 = 144$ is sufficient.

12. Hence we know that at the next level there are $2^{127} - 1$ or approx. 10^{38} dcss, and the perms to characterize them will be described using 256×256 matrices. They therefore span a space of dimension 256^2, and there is no hope at all that the perms will generate a full set of dcss at the next level. In the brief treatment we said that the construction broke down once you had found, in all, 137 matrices. However from our more careful point of view it is at the next stage that it breaks down. One can find 10^{38} perms but they will not generate more than 65536 dcss. This ends argument 2.

After the discovery of the construction in argument 2, considerable effort was put into deeper understanding of the mathematical system involved. That very interesting mathematical results were discovered was due in large part to John Amson. In time it became clear that what was missing, and so what made the construction hard to understand was the physics rather than the mathematics. The other approaches to the hierarchy model which we present start with more reference to experimental interpretation. The model in argument 1 follows the philosophy of quantum physics to the extent that it incorporates the interaction of the observing system within the physical process that is being observed. Nothing can be said about the physical world which does not start from this interaction. Much has been said in this book about the conceptual confusion that has arisen wherever efforts have been made to formalize the essentials of the observation process in the quantum theory, even though the insights of the founders of the theory seem to require this formalization. The confusion always seems connected with the unconscious intention of the quantum theorists to have their cake and eat it too by demanding that there should always be a background space for the quantum systems to be fitted into. We have already proposed our way out: we shall *construct* any space that we need.

Argument 3. Counter-firing

We take high energy scattering as the paradigm case of experimental knowledge, and indeed of any knowledge, even though an explicit policy of tracing all physical law to knowledge obtained this way is of recent origin and remains largely programmatic. We owe to H. Pierre Noyes the principle that counting of scattering processes in high-energy physics is primary. We do not allow ourselves to bring to the investigation of these primary phenomena a framework of physical law, or an array of associated physical concepts, which have had their origin in another sphere and nearer to traditionally more familiar experience. In the current quantum theory it is customary to give a measure of conceptual isolation to the "observation" or "measurement" by stressing that we lead up to it by an explicit 'preparation of states'. In this way an attempt is made to provide a primary unit of the kind required by Bohr's "quantum postulate". In our model, by contrast, we incorporate this insight into the formal structure of the theory rather than use it as a sort of verbal directive on the use of the mathematics. The following argument is given as a simple way to formalize this point of view.

We think of the firing of counters in an investigation of high-energy particle processes of some sort. We do not follow this course on a strictly operationalist prescription. That is to say, the theoretical concepts are not defined by these imagined experimental operations. It is only that it is convenient to imagine the mathematics with these processes in mind, realizing that in the quantum domain it will always be right, and always possible, to see the step-by-step nature of the model illustrated in reality because quantum physics is about the discrete aspects of the world and about the way the discrete things are investigated sequentially because of the nature of the experiments.

Let us suppose that the counters are wired up to lights on a panel. We know nothing of the wiring. We imagine the following course of events.

1. A counter fires. If there is only one call it no. 1. If several, label them in some way. Assume for the present that the system has an infinitely long memory and that therefore the labelling is retained until it is changed deliberately.
2. There is another firing. It may be the same counter or it may be a different one or different ones. Or it may be no. 1 and others together. If a counter fires, that is not the same situation as in 1), since 1) has by this time been labelled.

We could imagine a provisional notation for the various possibilities on somewhat these lines; first simplify matters by assuming that, at each step, at most one new counter fires:

$$1 \begin{array}{l} \to 2 \\ \to [1] \\ \to [1]2 \end{array}$$

and so on, the next stage being a choice of [1], [2], 3, [2], [2]3, 3[1], [1][2], [1][2]3, in the top and bottom branches, and a choice of [1], [2], [1] in the centre branch. Brackets have been used here to mark out a counter which has been labelled.

This provisional notation has a serious defect. In asserting that an identified counter fires again we have supposed that the firings are ordered, as though they were recorded by an observer who had an automatic sense of time, whereas we must remember that it is just this recursion back to previously established orderings which we set out to construct from more primitive notions. In addition it is clear that the notation will rapidly become too cumbersome to be useful. Some simplification (which is almost certain to involve rejection of some of the available information) is needed. On the other hand, this argument is useful in emphasizing the way the information about fundamental processes spreads out as we fill in more by constructing the processes outwards from something we know about.

Counter firing — more abstractly

To get a picture not open to these objections we have to look at things a little more generally. The essential feature of an experimental interaction — its developing quality — is what we plan to incorporate into the formal structure.

This is a novel feature for orthodox mathematics. We propose to do it by representing entities by finite but growing sets of elements. Thus a primitive time makes an explicit appearance in the mathematics: time, that is, as an ordering without any of the quantitive features engendered by clocks. This is evidently reminiscent of the thinking of Brouwer's intuitionism, but we are not arguing here for the adoption of intuitionistic mathematics. Brouwer's views on mathematics start from a different point from ours. The distinction between what we know of a process and the intervention or interaction through which we come to know it is provided by a partition or bisection of the finite set into disjoint parts. Let us introduce a little notation. If A, B are two disjoint sets of the finite but growing set S, we use $[A/B]$ as a new entity of S. This is one way in which S is growing, but there may be other ways. Another way to see the importance of the bisection is to try to formalize the idea of breaking into an experimental process. Here A might be the set existing before breaking in. (More correctly, to avoid ambiguity about an elements 'at break-in' we could say that A is the set not existing after breaking in, etc.)

Let us step back to reconsider our starting point. It must always be possible to represent what we know of the process as a step-by-step working back along the path which the events in question must have followed, so that our knowledge comes to be seen as built up of just such interactions as we first set out to formalize. Hence there is a basic recursive aspect to the model.

By a convention that is deeply rooted in our thinking this recursion is thought of as giving us information about persistent entities or particles, but any such idea can only be assimilated to our approach insofar as we can derive it as a consequence of the recursive method which we are trying to formalize. We shall in fact find that the definition of particles of this kind requires an initial specification of them through a variety of *descriptors* which appear in the technical setting as quantum numbers. The continuum ideas, with their associated dynamical concepts such as length, energy and momentum, are not in the natural path of the early development of our model, and we have to be content to see them emerge only gradually. A real task, therefore arises for us to tackle. We have to create the quantum numbers combinatorially, so as to explain why they exist in the form they do. Of course the classical definitions will afford guidance but ultimately we have to be independent of them.

We now use these ideas to look again at counter firing, by means of the notation which mirrors the generation of new elements by a bisection process.

Definition 5.1. If A, B are two disjoint (finite) sets of elements, then $\{A|B\}$ is an element, and all elements are got in this way. Thus in the argument of the counter, the occurence of $\{A|B\}$ might indicate a firing, in circumstances where B would be the set of already labelled counters that fired in this firing, and where A would be the set of already labelled counters that did not fire. $\{A|B\}$ is then defined as the bisection of the set $A \cup B$. The initial firing would now be $\{\langle\ \rangle|\langle\ \rangle\}$, where $\langle\ \rangle$ denotes the empty set: $\langle\ \rangle$ will be written "0". It is the existence of the empty set that allows the wholly recursive definition.

This procedure, and the basic idea of Definition 5.1, are to be found, used for rather different purposes, by Conway,[3] *On Numbers and Games*. We are indebted to this book for numerous technical points in what follows, though there are significant differences between our use and that of Conway, both in content and in interpretation. It is worth pointing out that the provisional interpretations in Conway's book runs into just the same complications as does the notation with brackets. When this alternative approach to counter firing is elaborated, it is found to give the very same mathematical system as those of arguments 1 and 2. This ends argument 3.

It is tempting to recur to the idea that we might specify our model operationally in terms of the counters. However macroscopic procedures can only be useful as a clear way of stating the properties which we postulate of the underlying entities: they cannot tell us what to postulate. In particular there is no macroscopic analogue to our use of the empty set, which therefore takes on the character of an arbitrarily chosen logical device. It is a device which is altogether the cornerstone of our model, and we are therefore bound to try to see it as a property of our fundamental entities, and no longer as an arbitrary postulate. Whether or not this attempt is plausible as an assumption about the real world (as distinct from being justifiable because its ultimate consequences work) has to be decided by considering what is implied in representing the existence and non-existence of entities by two-valued symbols, and in the possibility of developing a logic on that basis.

The only comparable step in the literature is that taken by von Weizsaecker who starts with primitive two-valued "Ur"s. We take that same method further than he in that we propose a logic in which the concatenation of the symbols is part of the underlying structure. Moreover our extension incorporates the interaction of the observing system, and therefore the quantum-mechanical philosophy, at that early stage, as a direct consequence of the extension.

It seems likely that one can only obtain a synthesis of these basic elements by adopting a throroughly constructive approach, as we have done. Some further backing is given to this conclusion by the early history of the model. Formerly, the elementary operations of the model were given a meaning by supposing that they represented interacting entities which could discriminate amongst each other in such a way that the discrimination between two entities represented by symbols would give a null if and only if the two new symbols were identical. One could see this interpretation as a way of representing the effects of the observing system and therefore as a realistic form for an 'observation logic'. However we shall show presently that the property of discrimination is a direct consequence of our constructive approach, and therefore that there is no need to argue for the two ways of regarding observation separately. In fact our present constructive underpinning of the older approach — apart from being a complete representation of the logical operations in a way the older one was not — gives a better picture of the observer philosophy of quantum physics.

Argument 4. Limited recall

This argument is intended to encapsulate the idea that we can only go back a certain number of steps into the past to identify the components of any quantum-type process with two inputs and one output.

Definition 5.2. A *system of limited recall*, LR(n), consists of a finite set of elements, $[a_0, a_1, \ldots, a_{n-1}]$. These elements receive *labels* through the following recursive process: If two elements a_i, a_j are presented they produce a third one. If they have already been labelled, the label of the resulting element, a_k say, is $f(a_i, a_j)$, where $f(a_i, a_j)$ is defined as the least label distinct from all $f(x, a_j)$ where x is any a_r with an earlier label than a_i, and from all $f(a_i, y)$ where y is any a_r with an earlier label than a_j. This construction, which is based on an idea of Conway (loc. cit.) is simply a formalization of the rather natural requirement that altering either of the inputs must alter the output.

We shall assume for the present purpose of getting to the first stage of the interpretation as quickly as possible that we can distinguish between an OPEN LR, simply denoted by LR, in which the cardinal of S is not specified and may be finite or may be enumerably infinite, and a CLOSED LR as in Definition 5.2 where S has exactly n members; and the latter we write LR(n).

Now the labelling is evidently unique, so that if the label of a_i is r and that of a_j is s, we can define a function Ω in the obvious way:

$$\Omega(r, s) = f(a_i, a_j).$$

It is clear that:

$$\Omega(0,0) = 0, \quad \Omega(0,1) = 1, \quad \Omega(0,2) = 2, \ldots$$
$$\Omega(1,0) = 1, \quad \Omega(1,1) = 0, \quad \Omega(1,2) = \text{etc.}$$

In general we can prove

Theorem 5.1. (a) Ω is commutative and associative, so that it is more natural to write

$$\Omega(b,c) = b + c$$

and 0 is then the zero element.

(b) The system of limited recall is then a fragment of Conway's field On[2] (that is to say $a + b$ is "nym-addition").

(c) Objects of S can be classified as being either "extenders" or "fillers". The extenders have to be imported without coming up in sums: the fillers do not. The distinction between extenders and fillers takes into account the fact that closure has to be imposed from outside. The generation process produces a new element whenever an extender is needed, whereas fillers are generated by the + operation.

From (b) of Theorem 5.1 it is obvious that the operation + in a closed recall system LR(n) of fixed n can be represented by the symmetric difference operation (addition mod 2, which we called discrimination in Argument 2). on strings of 0's and 1's — the operation being conducted elementwise. For example, any element such as 13 has to be written (in this case) as

$$8 + 4 + 1 = 1 + 0 \times 2 + 1 \times 4 + 1 \times 8,$$

and so is correlated with the vector (1, 0, 1, 1,). We shall call any binary associative operation with the property that, if and only if $a = b$, then $a+b=0$ for all a, b, a DISCRIMINATION. The reason for this name for this operation is that it can "discriminate" a pair of equal elements from a pair of unequal elements as we explained in Argument 2. Or — to put the matter in a way more appropriate to the use to which we shall put the idea — if we are given one element and we bring another to it, then the operation will tell us whether the new one is the same as it or different from it.

Definition 5.3. A set that is equipped with the operation of discrimination will be called a DISCRIMINATION SYSTEM.

We now get at once from Theorem 5.1

Theorem 5.2. (a) Any closed LR(n), with a fixed n, is a finite discrimination system, with vectors of length $\log(n)$ to base 2 (so that n is a power of 2).

(b) Any LR is a discrimination system.

It will be shown later that the converse holds, and that any finite discrimination system is isomorphic to a closed LR(n). This isomorphism requires n to be fixed once for all. Also, an enumerably infinite discrimination system is isomorphic to some LR. This concludes Argument 4.

It is convenient to break into the sequence of arguments at this stage to deal again and more fully with another concept which turns out to be important, in one form or another, in each of them. It is that of the DISCRIMINATELY CLOSED SUBSET (d.c.subset, or dcs.). This was defined in Argument 2, but we repeat the definition.

Definition 5.4. A d.c.subset T of a discrimination system S is a set of elements of S such that, for any two different elements a, b of T, $a + b$ belongs to T. The definition implies that the d.c.subsets are just the subgroups with the zero element removed, but this is not the most helpful way of looking at them.

To appreciate the significance of the d.c.subsets we notice that whatever the details of the interpretation of the mathematics turns out to be, we shall certainly find ourselves using the discrimination process to represent changes that come about in particle processes. Moreover the ordered sets of yes/no symbols that compose the participants in this process act elementwise, so as, even at this early stage to suggest the representation of a particle in terms of a set of all-or-none descriptors which could be an idealization of quantum numbers. Hence, supposing that the discrimination process has to be some kind of idealization of a scattering process, we can see that real importance must be attached to closure under the operation. One could assume some kind of ergodic property for a statistical background which would assure us that in the course of time all the members of any given d.c.subset would be generated, without its being necessary to make any special assumptions about the type of sequential operation which in fact took place. In this way one would be led to expect that the d.c.subsets would have a corresponding universality and independence of circumstances. By a continuation of this rather sketchy argument one would expect that the fixed points around which we are building our recursive structure — namely the d.c.subsets — would take their interpretation from the labelling scheme that we use for particles.

In earlier presentations of this work, the discrimination operation was imagined to be all that was needed to explain what was going on in the physical world as far as the model was concerned provided that a statistical hypothesis could be used. In fact the thinking had to be eked out with a physical assumption to the effect that the d.c.subsets established points at which the recursive generation process could be imagined to pause. It was supposed that identification with particles and/or quantum numbers would make it possible to attach labels to these pauses or development stages.

At that time we imagined scattering in terms of persistent entities which could be thought of individually. They were not thought of completely classically, but rather as things like monads which were each capable of reacting to another to the extent of affirming that the other was, or was not, the same as itself. That was all that they were capable of. In the former case the result of the discrimination was the null element: in all other cases an element of the discrimination system was generated which was different from both the originals. If it was not one already in the discrimination system at that time then it was added to it. (In the former case the new element was the null element, and was non-existent.) The natural appearance of the d.c.subsets in this picture has been explained. We see that vectors under discrimination were imagined to represent idealized particles directly — an interpretation which has since proved naive.

The d.c.subsets were now given a further role. The set of mappings which preserves a given (closed) d.c.subset (that is to say which maps each element of the subset into the same element, and does not do this to any of the elements outside the subset) is conveniently represented by a matrix operator which will be defined presently. There are two things to be realized about these mapping operators. Firstly they should have the same prominence in any interpretation that we have been led to assign to the d.c.subsets. Secondly they stand in a different and new logical relationship to the original elements of the discrimination system — namely as operators upon them. A combination of these two ideas led to the new vision of the mathematics of discrimination as being essentially hierarchical. The operators had to be new constituents of the universe in their own right, in any interpretation. Indeed it was only in this way that it could be possible to increase the size of the describable universe. One was led to the concept of the points, or elements, of the structure existing at different levels of increasing complexity, even though the points at the different levels were essentially the same sort of thing. A given point or element played dual roles according as it was an element in the counting process of the universe on the one hand, or was an operator specifying a set of steps in a set of such processes, on the other. In the first case one would be, as it were, taking part in the process; on the other, seeing the whole thing from outside.

It was this dual role that provided the characteristically quantum-theoretical 'observation logic' or interaction logic which we proposed at the beginning of this chapter as the aim we should have in setting up a mathematical model: and later improvements leave this part of the picture substantially unchanged. The way the interaction logic was achieved brings to mind the criteria set up by Bohr as conditions under which his 'quantum postulate' might be incorporated into physics. Petersen[4] (Chap. 3) quotes Bohr[3] (Chap 3) as follows: "For describing our mental activity we require, on the one hand, an objectively given content to be placed in opposition to a perceiving subject, while on the other hand, as is already implied in such an assertion, no sharp separation between subject and object can be maintained, since the perceiving subject also belongs to our mental content." And Petersen comments, "In other words, a separation is required to delineate an objective content, but the separation line is not fixed but moveable so that what is in one case on the subject side may in another case be part of the objective content."

In noticing the remarkable similarity between this description of what Bohr saw to be required for quantum mechanics and what we find it necessary to propose, we have to remember that Bohr was seeking to impose his view on a formal theory to which it was alien. It is probably for this reason that Bohr uses expressions — in common with other commentators — that imply so strongly the need for the interacting system to be human. He does this in defiance of the obvious objection to such forms of idealism that we really cannot think that the quantum universe would not work if there were no human beings. A Berkeleyan metaphysics at this point (stopping short of God and quads, though) may have been useful in insulating Bohr from the demand that he find the duality that is expressed in his view emerging from his basic formalism. Bohr's position is considered at greater length in Chapter 3 of this book, and in the final chapter it is pointed out that that there is no fundamental reason why the entities that play a part in the interactions should not be sufficiently complex to be human beings. However human beings are certainly not necessary. The same caution in comparing our ideas with those of Bohr should prevent us from leaping to the conclusion that our dual function of operators in the hierarchy algebra is the source of the complementary descriptions that form a basic part of Bohr's understanding of the quantum theory. There is indeed a connection between the ideas which is profound, but to tease out the complexities of the connections will take time.

Argument 5: Self organization

Our constructive point of view is of the same general kind as that encapsulated by recursive function theory, or, equivalently, by that line of thought that culminated in the 'Turing machine', and we shall compare our hierarchical structure with that. The detailed development of a hierarchical structure with a physical interpretation will have choices made at many points which will determine its form constructively. The decisions at these successive choice points will mirror what is going on in the world: we imagined them earlier as results of particle counts. Now a similar situation arises in discussing self-organizing systems. A Turing machine which could be called self-organizing would be one which changed its program according to its state. Such a Turing machine must operate, if we accept Turing's hypothesis or equivalently Church's thesis, just like a certain larger fixed-program Turing machine. The original machine was only putting up the illusion of being a self-organizing one. The usual way of avoiding the resulting limitation is by introducing a random element. Such an introduction of probability is what we have already presupposed in our model without discussing the reasons for it. We are really saying that the Turing machine model is not an appropriate one for a self-organizing system. A different theoretical structure is needed. Since Turing machines are simply a model for classical mathematics, this, in turn, is called into question.

The self-organizing aspect can be discussed in any model but we can conveniently refer here to argument 4 — the systems of limited recall. The self-organizing aspect of this entails a hierarchy with an interaction between levels, rather than a single system. This imbalance is rectified by an *economy process* in which certain special sets of locations can be given a single address without disturbing the discrimination in the minimal addressing procedure. Suppose that T is any set of non-null addresses, and define the *discriminate closure* of T, call it DT, recursively as follows:

(a) T is included in DT,

(b) If B, C, are any two different members of DT, then $B + C$ is a member of DT.

Note that this form of definition implies that any d.c.subset V is of the form $V = DT$ for some T. This makes a closed discrimination system have the form $S = \{0\} \cup DT$, for some DT. Consider now a mapping $\langle\ \rangle$ of a closed discrimination system, S, onto itself, which preserves the discrimination:

$$\langle\ \rangle : S \to S, \quad \langle\ \rangle(A + B) = \langle\ \rangle(A) + \langle\ \rangle(B).$$

Then, from Theorem 5.2, there is an evident representation of ⟨ ⟩ as an $n \times n$ matrix over Z_2, and since such matrices constitute a vector space of dimension n_2, the set of all such ⟨ ⟩ corresponds to a new closed discrimination system, $S * S$, say. This was discussed at some length in Argument 2. We note that the vector space picture suggests that ⟨ ⟩ is an element of different logical type from the addresses A, B, but in fact this difference is an illusion produced by the special representation. We may use $S * S$ for the economy process as follows:

Let DT be any d.c.subset. Then it can be proved that there exists at least one mapping ⟨ ⟩ with the property that

$$⟨\ ⟩(A) = A \leftrightarrow A \text{ is a member of } DT\ .$$

If also ⟨ ⟩ is chosen (as it can be) so that

$$⟨\ ⟩(A) = 0 \to A = 0\ ,$$

((⟨ ⟩ non-singular), we may use the new address ⟨ ⟩ to represent the old set of addresses DT, so that the information present at one level of the system is presented in a more economical way at a higher level. If, moreover, we choose the ⟨ ⟩'s for different d.c.subsets to be independent (that is, so that any k such themselves generate a d.c.subset of $2^k - 1$ members, and no fewer) then these ⟨ ⟩'s will serve to allow the whole process to be repeated, so that we will have a hierarchical structure just as we supposed earlier.

To summarize our discussion of the construction of a hierarchy we can speak of the "generation" of a hierarchy as the result of successive applications of the generation, equivalence, discrimination, economy rules in any order. Then we may call the generation of a hierarchy *complete* when it so happens that the creation and discrimination operations have been carried out in such an order as to maximize the information-carrying capacity of the structure. The complete hierarchy then serves to define the bounds on the amount of information that can be so dealt with. We can then prove (in line with Argument 2.)

Theorem 5.3. There is a unique complete hierarchy with more than two levels; it has successive completed levels of 3, 10, 137, $\sim 10^{38}$ elements, beyond which further extension is impossible. The derivation of the bounds of the levels, and of the stop-rule to the generation process at level 4, though not the demonstration of the existence of the hierarchy, is due to Parker-Rhodes.[1]

The proof of Theorem 5.3 is lengthy, and at present cumbersome, and is deferred to Chapter 6. For the purpose of this chapter in which the ideas are presented it seems better to indicate how the construction process proceeds at the lower levels and how the construction process has a stop-rule. A few remarks on the general significance of the theorem seem in order first.

As stated above, the theorem makes explicit how something less than the hierarchy can be generated, with the hierarchy serving as a bound on possible generation. Thus we imagine physical processes generating terms in the hierarchy, but not, in general, the whole of it. However the complete hierarchy is a kind of union of all those generated, and so provides a bound. This bound is the ultimate origin of the finiteness which characterizes the world of quantum physics.

In the mathematics leading up to Theorem 5.3 and its proof, exact form has been given to what was earlier described as an ergodic hypothesis. We are assuming that if we wait a reasonable time watching the events which appear under the conditions which correspond to a a given d.c.subset, then all the elements of that subset will manifest themselves with a probability which we can make as high as we like if we wait long enough. The way is now open to us to use the structure which we built to deal with circumstances where it is not the case that all the possibilities are realized. All the cases where there is an interesting diversity of structure are of this type, but it is important to realize that without the 'exoskeleton' provided by the limiting or bounding case of the complete hierarchy we could not describe the detail. Moreover this mathematical and logical fact has its counterpart in the fact of ordinary physical theory that one always uses a familiar range of physical fields or interactions (depending on how we care to speak) to describe the detail of a physical situation, whatever that situation may be.

To give a brief survey of the hierarchy construction which is yet a little fuller than the sketch in Argument 2., it is convenient to have an abbreviated notation for vectors and operators. We write, for any vector v, $v = i+j+\ldots+k$, where the only rows occupied by 1's are the ith, jth, ... (If we are in a low dimension so that none of $i, j, \ldots$ exceeds 9, then we simply write $ij\ldots k$.) An operator can then be written as an assemblage of column vectors, and this notation allows calculations to be carried out quickly, since

$$(P, Q, R)1 = P, \qquad (P, Q, R)2 = Q\,.$$

One could regard this formulation as a different (though equivalent) calculus to matrix algebra over the field Z_2, and it is useful to see that various such schemes are possible given that certain minimal conditions obtain. To begin this repetition of the construction, choose two basis vectors in two dimensions, 1, and 2. Then we have to find non-singular operators with the invariant spaces:

(a) {1}, and this is evidently the operator (1, 12).
(b) {2}, and this is (12, 2).
(c) {1, 2, 12}, and this is the unit matrix, (1, 2).

It should be noted that there is no choice at this stage. These operators may be written at the next level as vectors, say as 134, 124, 14. For some purposes it is necessary to keep them in this form, but for the mere existence theorem it is possible to simplify by taking these as a new basis, 1, 2, 3. There are now 7 invariant subspaces, and it is possible to find 7 corresponding non-singular operators which are linearly independent in a number of ways. For example, in the three-dimensional subspace, the operator (1, 3, 23) has unique eigenvector 1, and so, interchanging the first and second vectors together with the places where they are written in the operator, (3, 2, 13) has 2, and (12, 1, 3), has 3. In much the same way the three (13, 2, 3), (1, 23, 3), (1, 2, 13) serve for the three element spaces, and (1, 2, 3), the unit matrix for all 7 vectors. It is not hard to verify that these are all linearly independent.

It is harder to establish the existence of the 127 operators at the next level, but several different versions of this have now been carried out. The termination of the process, as we pointed out in Argument 2., arises because the dimensionality of the space does not increase fast enough to accomodate the number of linearly independent vectors, as we see from Table 5.1.

Table 5.1.

Number of vectors	2	3	7	127	$\sim 1.7 \times 10^{38}$
Dimension	2	4	16	256	65536

It should be borne in mind that it is the bounds to the hierarchy levels that are the first theoretically significant numbers to appear in the theory. They are the total numbers of discriminations whose cardinals have already been calculated. These cardinals are in fact the cumulative sums of the numbers of vectors given in Table 5.1. It is helpful to remember that generation of the discriminable entities at a given level cannot take place until those at the simpler level all exist.

We conclude this chapter with one further argument, that has been used to short-cut the rather elaborate generation of the hierarchy that is to appear in the next chapter, by means of a computer model.

Argument 6. Program universe

This computer model originated in the search for the simplest way to generate strings of binary digits (0's and 1's) which could have an immediate identification as discrete forms of the conventional dynamical concepts (spatial configuration and momenta and the rest). There are two operations in the program, called respectively PICK and TICK. When there is a collection of N strings in the memory, PICK selects two of them and discriminates them; then it tests whether the resulting string is already in the memory. If not, it is adjoined and the program moves on to PICK. If it is in the memory, the machine moves on to TICK again. This adds an arbitrary unit to each string in the memory, so that the common length of all strings increases by one and the system returns to PICK. (The system begins with no strings and a slightly modified version of TICK in which two bits are generated arbitrarily.)

An important result in PU (Program universe) is that TICK results from either what Noyes calls a 3-event or from a 4-event. It would be better to call these 3-processes and 4-processes. Immediately before a TICK there are three strings a, b, c, satisfying $a + b + c = 0$. These are replaced, at the TICK, by a/t, b/u, c/v where t, u, v, represent the extra bit added at the end of every string by the operation. In one half of the possible cases $t + u + v = 0$, so that $a' + b' + c' = 0$, and if these are picked again, we have a 3-process. In the other half of the possible cases, at the next PICK, $a' + b' + c' = d = t + u + v$, and $a' + b' + c' = a/t + b/u + c/v = 0/1$ (say) which Noyes calls a 4-process.

So far the PU scheme is simply a way of generating strings, without the structure which we have been discussing in the other arguments. This structure is moved by PU into the labelling routine for strings. Once two linearly independent strings have been generated and labelled, this is called Level 1, and the next strings — once three more linearly independent ones have been found — form the basis from which level 2 can be generated by discriminate closure, and so on. Although this way of looking at things misses some of the structure of the earlier examples, it does produce an important insight. After the fourth level has been generated, when linear independence is no longer possible, it is none the less possible to continue to generate strings, and so longer and longer strings can result. Each longer string divides into two parts, which may be called the initial and the continuing parts. The initial part, which is finite, is the part which takes part in the hierarchy construction. The continuing part has no stop rule and so can be of any length. These two parts are called by Noyes the 'label' and the 'address', but it seems more in keeping with computer terminology to call them respectively the 'label' or the 'address', and the 'content'. (The content is the variable part.) The idea behind this distinction was that labels would be interpreted as quantum numbers and content would cover the variable attributes of particles like momentum, in scattering processes. This interpretation is now seen as naive, and illustrates the limitations of PU, but was meant to work as follows.

The physical interpretation of PU is provided by identifying a 3-process with a scattering 3-vertex. At such a 3-process, if one writes p_a for the number of 1's in the string a (and so for b, c) then it is clear that, since in discrimination the number of ones cannot increase but may decrease, we have $p_a + p_b \leq p_c$ and two similar inequalities. These triangle inequalities are sufficient to ensure that the quantities p_a etc. can be regarded as the magnitudes of three vectors adding up to zero. The continuing part of the strings are then interpreted as describing the momenta of the particles taking part in the scattering process, whilst the finite parts are interpreted as the specification (in some suitable way) of the particles taking part in the scattering. This specification takes the place of the quantum numbers of the orthodox theory, and is, in fact, fairly closely connected with it. The success of the scheme is to establish that there must be entities with an exhaustible set of attributes, which was previously mysterious, and to unify these attributes with that of momentum, which may have many values (in the orthodox theory, a continuum of values, but here a discrete but dense set).

Some features of each of the six arguments of this chapter will reappear in the rigorous version of the mathematical theory which we shall formulate in Chapter 6.

CHAPTER 6

A Hierarchical Combinatorial Model
— Full Treatment

The arguments of the last chapter pointed in the direction of a hierarchical combinatorial model, but all had limitations in the shape of places where the mathematics had to be eked out by analogies of varying plausibility. This was because the models tried to interpret existing mathematical calculi. In this chapter we shall give a theoretical analysis of the process. This theoretical analysis takes the place of what was called "mathematics" in Chapter 5. That term is misleading insofar as it implies that there is a calculus which is then interpreted in terms of some independent physical world which we can describe without the model. The theoretical analysis develops as the process goes on. Its "mathematical" quality lies only in the use of symbols in a way familiar from algebra. We begin with the process model, but first we give a short statement of what it describes — a description of the kind of universe it presupposes.

A process universe

To start with there is a division between the known part of the universe and what remains unknown. Information about the unknown part is to be obtained only through indirect statistical inference, and our theoretical structure relates to the known part. Changes in the division are described as being due to an *entity* changing from being in the unknown to being in the known part.

Such change takes place in discrete steps, and in chains or sequences. Such a change can be thought of as a universal basic observation process. It could in special (and of course very complex) conditions be conscious, though in general it will not be.

We are following the tradition of quantum physics in incorporating the observation into basic theory but we avoid an unbelievable distinction between those processes which have to do with observation and those which do not. The elements of the theory are accordingly sequences of steps which represent processes in the world. The simplest (and therefore the correct) way to begin is to imagine a step of a sequence as a decision or "discrimination" process which asserts or denies that an entity is identical with or different from another entity from the unknown part of the universe. If two such entities are equal, they belong to the same set, and so the discrimination steps tell us something important about the way structure can be built into our growing system by back reference.

From definition, every two entities are different. If pressed we have to say that these entities could be anything in themselves: the only thing we can say about them individually arises out of recognizing the entities which are repetitions of earlier ones. Such entities will be called *equal*, and the process can be seen as a grouping of equal entities into a set. This is important because the statistical inference mentioned above as the one and only one source of information about the world of the entities requires that we be able to represent the frequency of appearance of one particular set of them.

Before going into details, we sum up the position.

1. The way in which this chapter differs from the earlier arguments described in Chapter 5 is in taking very seriously the emphasis on the process taking place rather than on the objects involved in it. For example, the earlier arguments frequently referred to certain sets of elements. Here we shall be concerned less with the set than with the process of determining whether a new entity belongs to it or not.

2. Such a process theory must 'work by itself' in the sense that it must never be necessary for the mathematician to intervene. The extent to which this self-denying ordinance has been successfully obeyed here owes much to acute criticism by Alison Watson whose views[1] have influenced our thinking on the metaphysics of process. Only what is laid down in the principles (or prescribed by the program) may occur, and, even more important, all that is allowed by the principles must eventually occur. (This latter assumption is a sort of discrete ergodic principle, which is used extensively in Chapter 7.)

3. By definition nothing is known about the background world, and this ignorance is represented by the use of randomization. We do not mean that

things are 'in their nature' random (whatever that may mean) but that in the absence of knowledge to the contrary all possibilities must be treated indifferently. This principle (essential to us in making calculations) corresponds to a certain universality and absence of specificity which one presupposes when one analyses the world the way physics does.

4. The 'background universe' to which we have appealed has no space continuum, and all that we have assumed about time is contained in the principle of sequentiality. We assume that it will be possible to obtain discrete approximations to the continua from the only connectivity we have up till now — namely cross-referencing of sequences. However the road will be long. To reach this cross-referencing we have to turn to the theoretical analysis of the process which was mentioned above. We spoke earlier of grouping entities into sets of equal ones. To have this, there must be a notion of equivalence between two examples of an entity so that, for each potentially new entity it can be determined whether it is equivalent to one which has gone before. It is at this point that the context-independence of the system is ensured. The introduction of this particular form of equivalence relation (that is, a dichotomous choice between $x = y$ or not) excludes Parker-Rhodes' indistinguishables. Instead, something very like Leibniz's "identity of indiscernables" is imported — but with this important difference: Leibniz wanted to consider an object as its bundle of attributes, and such a definition might nowadays be very congenial to elementary particle physicists. But Leibniz got into difficulties because of contradictions about whether space and time were to be counted among the attributes. We pursue the same line as Leibniz but space and time do not at this stage exist in the system so that no such contradiction can arise.

The equivalence arises in a system that is being continuously generated so that the process of determining the equivalence cannot have the form of a fixed algorithm. We shall find when we carry out the analysis of it that one consequence is that, although we start with the developing sets that have been described, finite sets of these arise necessarily. Just as the mechanics of recognizing a single entity is all that we can say about it, so these finite sets are characterized by the mechanics of recognizing whether an element is in the set or not. The recognition process may then be used as a unit in its own right. This innovation is the real origin of the theoretical structure which has already become familiar from the arguments of Chapter 5. We call the structure the 'combinatorial hierarchy'.

These finite sets are a first indication of a more complex aspect of the process. We can set out the complexity that the process must have in three stages:

(a) The unambiguous determination of whether an entity belongs to an existing set or not.
(b) The treating of a set as a single element (both in the usual sense of set theory and as seeing the recognition process as a unit in its own right, as stated above).
(c) A third requirement which arises from the discrete character of the process. The entity has to be located in a definite but growing "field" of sets to which it might have belonged. This restriction is needed because otherwise the assigning of an entity to a set would correspond to an infinite amount of information, which would subvert the discrete character of the model which the notion of 'entity' was intended to enshrine.

As soon as a set of (one or more) entities has arisen in the process, an independent analysis can begin by introducing an element to correspond to the set and by employing a symbol, which we shall call a label, to denote this element. It will then be convenient to use the term "element" for the set of entities in the original process as well. It will be a convenient shorthand to say, "the process has labelled an entity with label 'a' " when what is meant is: the entity has come into the known part of the universe and has become unambiguously assigned to some set of entities by the process, and in our analysis we have introduced the 'label' to play the part of the set. That is to say, we find it convenient to use the label of a set of equal entities as a label of the entities as well; no confusion will result.

Labels are so called to distinguish them from other symbols which will arise in the analysis. Examples of such symbols are:

(i) a symbol loosely associated with an element, but not yet satisfying all the requirements (a), (b), and (c) above.
(ii) a symbol which might potentially be associated with an element, but that element has not yet arisen in the process.
(iii) a symbol which can never be associated with an element. The need for one such symbol will appear below.

The requirement (c) above shows the need for symbols of type (ii) but without showing how they are to be provided. This is a first indication of the need for a parallel but separately generated system, to which we shall recur later.

'Label' is a powerful metaphor which is intended to suggest interference from outside. (Something who or which does the labelling.) Either this interference is a human act or else we are talking about the development of a parallel and separately operable system. From the requirement 2 above, that the process must work by itself, the first possibility is ruled out. In classical physics it is implicitly assumed that the mathematician can direct his attention at will to any part of the physical process he wishes, and so the labelling is automatic and unconscious. Human interference is tacitly assumed. A theory is quite different, and outside interference is disallowed. Therefore it must adopt the second possibility of a parallel or separately operable system.

We shall later find that our constructive process generates a parallel system automatically and that system is of a sort which permits us to randomize over the choices so as to put off for quite a long time its essentially freely creative aspect. The current quantum theory stands with a foot in both camps, since, as we have extensively argued in this book, it cannot avoid appeal to changes which are produced by the observer. The notion of label is strictly part of the analysis of the process which we are carrying out. It is a legitimate part exactly to the extent that it captures the the aspects (a), (b), (c) of the process and imports no more.

To have a concrete example to hand, one can think of the labels as the "bit-strings" of the last chapter.

$$a = (a_1, a_2, \ldots, a_n), \qquad b = (b_1, b_2, \ldots, b_n),$$

namely strings of 0's and 1's with discrimination between a and b represented by

$$a + b = (a_1 + b_1, a_2 + b_2, \ldots)$$

the addition being modulo 2. It will transpire that this way of thinking loses very little in the way of generality except for the artificial restriction of a fixed value for n. The reader will notice that this particular example of a label does not provide symbolism for the only property that elements can have, viz. the frequency of their appearance. It provides only a means for their identification. It is impossible to provide the frequency at this stage since it is only relative frequencies that can arise and these cannot arise until the other elements are in play. As a first step towards these frequencies' appearing, we could extend any labelling system such as the one in the example to a set of ordered pairs, (a, n) where a is one of the labels of which the example was intended to be representative, and the natural number n is a tally of the number of times. This is still not enough for the frequency to be able to arise, since that will require something analogous to memory as understood in computing. In order to accommodate this, it will be necessary to embed the mathematical system which has arisen in some more general framework.

The Principle of Choice

So far we have described the process in very general terms and one may wonder how a detailed analysis of such a general system is possible. That it is possible is a consequence of the validity of a peculiar form of argument which will recur a number of times in our discussion and which we now describe. The validity of the argument will be self-evident and we shall refer to its use, for shortness as the Principle of Choice. The principle is as follows: in analysing some part of the process it will usually be the case that a number of different theoretical analyses describe it equally well. When this is so, the principle simply states that we can choose any one of these analyses to the exclusion of the others. We make such a choice on the grounds of convenience. The nature of the principle will be illustrated by the first application of it.

We should naturally like the labelling process to be systematic, in the sense that labels are constructed from a fixed set of symbols defined before the process begins. By contrast, the system of using girls' names to label hurricanes is not systematic: it is hopelessly limited and impossible to analyse. However, it is physically quite unimportant how the labelling is carried on so long as it is done in accordance with what has been said before, equivalent objects receiving the same label, subject to the conditions (a), (b) and (c) above. Such different procedures of labelling differ only in features of no physical significance and so we can choose to analyse any one of them, and we choose a systematic one. In future such a detailed discussion will be summarized by saying: "By the Principle of Choice the labelling will be assumed systematic".

By the Principle of Choice we may use the set L,

$$L = [1, 2, 3, 4, \ldots] \,.$$

The set is infinite, but this is not a fresh assumption. For example, this set could be coded in terms of the set [1,2] with two symbols by the well-known rules:

$$1 = 1, \quad 2 = 2, \quad 11 = 3, \quad 12 = 4, \quad 21 = 5, \quad 22 = 6, \ldots \,.$$

The symbols $1, 2, 3, \ldots$ are not, of course, the cardinal numbers. But we shall appeal again to the Principle of Choice to use them in such a way that they have an interpretation as ordinal numbers; the symbol m will have the interpretation of marking the mth step in the development of the labelling. The symbols of L are ordered by their ordinal character. This ordering can be extended to the labels as well in various ways. It is convenient to defer the description of the exact ordering process until later.

When L is employed as the set, the labels will be strings in $1, 2, 3, \ldots$. Here some care in notation is necessary. If we follow the usual practice, and write a string as (say) 2714, this is unambiguous as a label only if we are using no more than the first nine elements of L. Otherwise, 27 might designate the twenty-seventh letter of L instead of the string of 2 followed by 7. So it is necessary to write the strings of L in the form (2,7,1,4,). (For certain of the calculations which follow we can drop the brackets and the commas, and still avoid muddles, because these calculations take place at an early stage of the elaboration of the system and only the first few of the symbols of L are employed, and we shall take advantage of this.)

Let us summarize the position we have so far reached.

1. There is a set $[1, 2, 3, \ldots]$, and labels will be strings. (Notice that it is not stated that all strings are labels, nor that different strings are necessarily different labels.) We call the set of strings $W(L)$.

2. Entities become known in the process and this fact is analysed by assigning as a label, a string in L.

When a string labels an entity, the string is a label. When an entity has been labelled, it becomes an element of the system.

3. One label will usually apply to more than one entity. The corresponding entities are then equal.

4. Because of 3, operations on elements in the course of the process can be written down as operations on labels. This is the transition from what Brouwer called the Ur-Intuition, to mathematics.

5. The nature of the operations in 4) was not specified. If we now confine our attention to those operations which give further strings in L, these may or may not be labels. They will be labels if the operations correspond to the steps of the process. Other operations which arise as part of the parallel and separately operable system may or may not give rise to labels.

Testing

We now use this formal apparatus to investigate how the labelling is determined by the operation of the process. At any stage, suppose S to be the set of elements which have already been labelled. If a new entity becomes known, the subsequent operation of the system must be to test this against S. That must mean that the operation of the system relates S and the new entity in some way. The establishment of this relationship has to show up in the analysis. If the new entity belongs to one of the elements of S, then its label is determined. The only change is the increase in size of that element — an increase which calls for no change in notation since it is agreed that we are not dealing with the fixed sets of classical mathematics, but with a continually developing system. Thus no new label can arise in this case, and so the establishment of the relationship has to be indicated in some other way. By the Principle of Choice there must be a symbol to show that the entity under test is not really new, and this symbol cannot be a label. It must be of the type (iii) above. We call such a symbol a *signal*. By the Principle of Choice we can take this symbol to be unique. If this signal does not arise, the entity is given a new label. By the Principle of Choice we may take this new label to be the least. It is then distinguished from other entities and it is unambiguously determined whether or not the process is dealing with it at any subsequent stage of the construction. It is otherwise if the process produces the agreed signal. The entity is then not really new, so the process must go on to determine which of the elements of S is the equivalent one, so as to determine the appropriate label. This can be accomplished by testing the new entity against the various subsets of S.

Two questions need to be answered about the agreed unique signal. It is clear that it is needed, but in what way has anything been gained in using it to replace the growing set S of already labelled elements? Evidently something is gained by the increased precision. But where does this come from? The answer is that this is the first appearance of an essential asymmetry. It is necessary to have such an asymmetry before one can talk — for example — about a distinction between what is observed and the observing system. This is not in very precise terms; it only says that there has to be an asymmetry of function in that the action has to be initiated somewhere. There has to be a point at which we break into the system, and the most profound place seems to be to take the agreed signal inside the primary boundary so that the process builds up a structure to explore outside it.

There is, then, a unique signal for equivalence of entities, and it cannot be a label because no new entity has entered. This is an indication of the general coherence of the system, for if it were a label, it would be necessary to test whether the result of a test was this label or not, giving rise to an infinite regress. Because it is not a label, it is at once apparent whether two entities are equivalent or not. In such a system we shall write the unique agreed signal for equivalence as z for the present.

We said above that the set of labels was ordered. This is achieved by the slightly more general systematic ordering of all the strings in $W(L)$. We leave the details of the ordering for the moment because it will transpire that not all strings in L are to be reckoned as different labels. It is obvious that some systematic ordering can be given. It is convenient to adjoin the extra symbol z to this ordering by stating that it is less than all labels. Then in terms of this ordering there is no loss of generality in requiring that each label that is introduced into the system is the least one which is not precluded by its having been used before (or by any other restriction which may be introduced). When z is included we write the set of strings as $W(L, z)$.

Test functions

We now begin to develop a formalism on the basis of these definitions to describe the process of testing new elements. We start from the notion of an entity (not yet labelled, so unable to enter the mathematics at this stage) tested against the (developing) set S and giving rise to a signal. We do not wish to distinguish strongly between the set S and the process of determining its elements, so we shall use the same symbols for the process and for the set. As an initial notation, we could write this as

$$S \to a \quad \text{or} \quad S \to z$$

corresponding to the two cases of a genuine new entry or an old one. Here a is any label and its value is of no relevance. In subsequent testing it may happen that the same process S is used on an element which has been labelled, say b. (For example, it may be that it is necessary to determine whether a 'new' entity which turns out not to be genuinely new, gives the same result in a process S as b does.) Then one would have to modify the notation to show that b enters, and so write

$$S, b \to a \, .$$

The mathematical reader will notice the similarity of this situation to that of a function, which would more usually be written

$$S(b) = a \, .$$

This is actually a more convenient notation and we shall adopt it but we must make some qualifications:

1. The operation on the left-hand side is not a functional symbol in any sense which has become common this century. None of the refinements of domain or range or such concepts are possible here.
2. Leaving that on one side, a function is usually thought of as a precise rule which specifies a value (a) when b is specified. Our system is not yet precise

in this way. If the new entity is genuinely new, the process S may give rise to any signal whatever, except z.

Our mathematical reader may be thinking that the refusal to consider the equation

$$S(b) = a$$

as referring to a function is because the system is not working with functions but with equivalence classes of functions. That is, he would argue that one could call two functions equivalent exactly when they agreed about the elements to which they assign the value z. But this would be just a mathematician's device, because the process does not work with equivalence classes of functions but with the operations themselves. None the less, the mathematician's device does give a useful clue to how we may proceed. The process evidently works in exactly the same way for any one of an equivalence class of rules. It does not matter which one is being used. By further applications of the Principle of Choice we select particular (equivalent) forms of operation of the system for study because their mathematics is more transparent, knowing that the results we get are of general applicability. It will then turn out that the initially imprecise operation S can be given sufficient precision to have the form of a function, as we shall show.

The one element case

We noted above that when potentially new elements are tested against S and found to be, in fact, old, it is necessary for the process to test them against subsets of S in order to label them correctly. In particular, a continuing use of this argument will often lead to the testing against a set U containing only one element, u say. We shall look at this case in more detail. It will appear later that it is different in kind from the case of an S of more than one element. The one-element case determines the nature of the testing and therefore its investigation has to be carried out first. Suppose U, V, W are three sets of single elements u, v, w respectively. Then the testing of whether x belongs to U is exactly the same as asking whether or not x and u are equivalent. The investigation takes the form of several applications of the Principle of Choice to the conditions that the relation of being equivalent must satisfy simply in virtue of its being an equivalence relation. These form the first non-obvious applications of the Principle of Choice. The details are a little complicated and so we defer them to note 1 at the end of this chapter and proceed to state the result here. This result is most naturally expressed in terms of the concept of a row. A row is a string in L of the form:

$$r = (r_1, r_2, r_3, \ldots, r_k) \quad \text{where} \quad r_1 < r_2 < \ldots < r_k.$$

The order of the row above is defined as r_k, the greatest member of L and is written ord r. The norm of r, $|r|$, is the number of members of L in r.

The number of rows of order n is easily seen to be 2^{n-1}; that of order not exceeding n is therefore n^* where, here and elsewhere, we write n^* for the number $2^n - 1$.

To any string in L, w say, there corresponds a unique row, $r = \rho(w)$ constructed in the following way:

(a) Remove any pair of occurrences of a symbol of L,
(b) order the remaining symbols of L in ascending order.

Thus, $\rho(5, 3, 2, 3, 7, 1, 3, 5,) = (1, 2, 3, 7)$. A binary operation, denoted by $r + s$ is defined between any two rows r, s as follows: denote by $r.s$ the string in L formed by concatenation (r followed by s). Then $r + s = \rho(r.s)$. It is clear that $+$ is commutative and associative, which justifies the use of the sign $+$. No confusion will result from the use of the same sign as for discrimination in the last chapter because the present operation will turn out to be very closely related to discrimination.

In terms of rows we can now state, as a consequence of the properties of an equivalence relation:

Theorem 6.1. *The process works as if all labels are rows, and the test-signal for x to belong to $U = [u]$, where x, u are rows, is $x + u$.*

It is now easy to construct the operation $x + u$ recursively, and so to explain the initial stages of the process. The first label given to an entity is the least, i.e. 1. When another entity comes into play it is given the next symbol 2 and in the course of testing that it is new the signal $1 + 2 = (1, 2)$ is a signal. This signal is a label, for it labels a part of the process, i.e. the testing of 2 against 1. Then also $1 + (1, 2) = 2 + (1, 2) = 1$. Thus the three elements 1, 2, (1,2) are such that the result of testing any one of them against any one of the others is to give the third. This structure is that of the quadratic group, except that the identity element has been omitted. The element z is the omitted (zero) element, and so, from now onwards, we write 0 instead of z. The operation of testing, $u + v$, is called *discrimination*. This is an appropriate notation because the operation $+$ is commutative and associative. The quadratic group is the first non-trivial example of a discriminately closed subset.

Continuing the recursive process in the new notations, the next new element is 3. The values of $1 + 3$, $2 + 3$ come out at once. That of $(1, 2) + 3$ will be $(1, 2, 3)$ by the rules. Thus the set consisting of the first seven labels listed in the ordering, together with 0, form an abelian group of order eight, which is a direct product of three copies of the group of order two. The first seven labels thus form a larger discriminately closed subset. Since each new discriminately closed subset (dcs) consists of one new element together with the result of discriminating it with each element of the last dcs, it follows that the size of each is one more than double that of the previous one. So successive dcs are of sizes 3, 7, 15, 31, ... and every dcs is of size r^* for some integer r (with the definition of r^* above). Correspondingly, if it happens that the set of elements generated up to a certain point in the process is not a dcs:

Theorem 6.2. At every stage the system can be embedded in a dcs of size r^*, for suitable r.

Other notations

In the early accounts of the theory, some of which are described in Chapter 5, two equivalent or almost equivalent notations were used. We shall now show the connections with those earlier notations. The first one was the bit-string notation, or equivalently that of vector spaces over the field with two elements, Z_2. The relation with that notation is provided by:

Theorem 6.3. Any dcs can be mapped onto a vector space over Z_2, where discrimination is the addition operation in the space. If the dcs has size r^*, the space has dimension r.

The proof of this is easy. Any row, say (1,3,5), is mapped into the vector with a 1 in the first, third, and fifth places. The verification of the rest is immediate. It is important to bear in mind that it is only at some fixed point in the development, for a fixed dcs, that the theorem holds. As the development proceeds, the dimension of the vector space has to be continually increased. This may seem to be a flaw in the vector space picture — only now interesting because of its history. In fact the vector space picture, or something equivalent, is necessary. A larger logical structure in which the discrimination system can be embedded was mentioned as a necessity above in the discussion of multiple testing of an unlabelled entity. If the value of n is fixed one has available for the first time a static system in which the notion of *memory* becomes meaningful.

The second notation is that in terms of finite fields, which is used by Conway (loc. cit.). Here the relation is:

Theorem 6.4. Any dcs can be mapped into a finite field of characteristic 2, where discrimination is the addition operation in the field. If the dcs has size r^*, the size of the field is $(s^* + 1)^* + 1$, where the integer s is defined by

$$(s-1)^* < r - 1 \leq s^* \;.$$

The size of the field is determined not by the closure under the addition operation (which would have given a smaller size) but closure under the multiplication operation which a field must have (see Note 2 at end of chapter). The introduction of a second operation, multiplication, without physical interpretation at this stage, makes the finite field notation unhelpful as well. In the finite field case the mapping is most simply defined in terms of the ordering used above for rows. If the elements are written out in ascending order and numbered off, the elements numbering them off (the ordinal numbers) will be the corresponding field elements. Thus the correspondence is:

1,	2,	(1, 2),	3,	(1, 3),	(2, 3),	(1, 2, 3),	4,	...
1,	2,	3,	4,	5,	6,	7,	8,	...

The sizes of the successive fields, given by Theorem 4, are:

$$2,\ 4,\ 256,\ 65526, \ldots ,$$

each being the square of the previous one. It will transpire below that there is a physical significance in the introduction of the second operation, and so, correspondingly, these numbers are of more universal applicability than the derivation here suggests. For the present, we shall stay with the notation of rows, with reference to the vector-space or bit-string notation only for historical comparison. Let us summarize the position reached up to now: we have shown that a process theory of the sort described in the introduction to this chapter will consist of steps in a sequence. Each such step is a discrimination between two entities which determines how these entities are to be labelled. When they have been labelled, and so become elements in the mathematics, these elements will be rows. These may, if desired, be thought of as bit-strings. The only limitation is that they are of indefinite length.

The n-element case

The notion of discrimination which was formerly somewhat ad hoc has now been elucidated by looking in detail at the particular case of testing a new element against a single one. It is this case which has to be treated first in order to formulate the concept of discrimination. It is now time to return, with the results now gained at our disposal, to see how the general process will deal with the general case of testing against a set of already labelled elements. There are two related difficulties which have to be faced in such a procedure. The first one is that the set S is not a fixed set but is growing as the process goes on. The use of the finite signal set Z (now specialized to have only one member, 0) is the device that deals with that. The second difficulty — that of determining whether a new element x belongs to the set $S = [u_1, u_2, \ldots]$ of already generated ones — cannot be answered by simply requiring that each of the u_i be tested in turn, for such a procedure needs to ask at each stage "Has this u_i been used before or not?" And this question can be answered only by a further discrimination, and so on in an infinite regress.

Instead, the system must contain a rule, as in the case of testing against a single element, which provides a signal (which can be taken to be zero at once, since the preliminary argument is like that of the previous case) exactly in the case when any of the $u_i + x$ vanishes. That is to say that the procedure must relate S and the new entity x so that (again using the same symbol S and for the process of determining its members):

$$S(x) = 0 \quad \text{if and only if: one of the } u_i + x = 0 \text{ ,}$$

but it must do this as a whole, not using one member of S at a time. Exactly as above, S is not a function (yet) for its value is not prescribed except at the finite set of arguments in S. Its value for other arguments is quite arbitrary, but we can short-circuit the rather lengthy argument which we used for the simpler case when S had only one element by remarking that the same considerations apply here. Therefore we can make the definition of the operation S more and more precise while still leaving the action of the process exactly as it was.

At this point the reader may well enquire whether, because of the form of the condition above on S, we cannot simply rely on the equivalence of the process to a finite field, in which there is then a multiplication operation (denoted in the usual way by juxtaposition) such that $xy = 0$ only if x or y is 0. The answer is that this is a particular way in which the process might operate, and would indeed be equivalent to the ones which we are going to describe. The objection to it is that it has been introduced arbitrarily, without physical justification. As a consequence it does not tell us as much about the process as we shall be able to glean from the alternative which we shall follow. The way in which multiplication would be used would be to write S directly as a function:

$$S(x) = (u_1 + x)(u_2 + x) \ldots \text{ ,}$$

although this arbitrary procedure does not tell us as much as it might, it does point to the numbers 4, 16, 256, 65536, ... which derive from the closed sets under multiplication. These numbers are indeed of physical importance, and will appear again in our discussions.

A hierarchy of levels

Suppose that the operation of testing against S has been made precise in the form of a function, for example, as in the case of using multiplication. Then if S, T are two sets of elements, it is possible to define an addition operation between them by the rule:

$$(S+T)(x) = S(x) + T(x)$$

for all x in play up to the point at which the definition is used. Then it is easy to see that this operation has exactly the same properties as the original discrimination operation. For example, one defines $S = T$ to mean that $S(x) = T(x)$ for all x, and one defines $S+T = 0$ to mean that $S(x)+T(x) = 0$ for all x. Then one can see that $S+T = 0$ if and only if $S = T$. It is important to notice that the way of combination of S and T here is not related to the usual set operations. This is because it is the processes by which set membership is ascertained that are being combined, not the sets themselves. This is in keeping with what we said at the beginning: it is the process which is under discussion not the objects. As a consequence, if the operations of the system are made precise, so as to be functions, the set of such functions is itself a new discrimination system. The theory therefore automatically has a system of levels. The system is a means of talking about the process-oriented aspect of the universe. The labelling of elements in this discrimination system depends on the functions, and now these functions are themselves seen as part of a discrimination system but on a higher level, as it were. The original system has functions corresponding to sets of elements; these functions are the single elements corresponding to sets of elements at the lower level. Whether the system changes level at any point is not laid down. It may continue to treat functions as grouping elements at the lower level for more or fewer steps. As we explained at the beginning of the chapter, all that is allowed by the principles must occur; the various possibilities must be treated indifferently, and so the change of levels has to be considered on the same footing as staying at the same level. We remarked that the elements at the lowest level could not be made very precise. (We said that the elements could be absolutely anything in themselves; we can only talk about the mechanics of recognizing them.)

But those at higher levels are sets of elements at lower levels; so that more is specified about them in themselves than at the lowest level.

We now find an explicit form for the precise functional forms of the process, which we shall call *characteristic functions*. Just as with the basic case of a set of one element only, which affords a guide in our extension of the construction, there will be many functions, any of which do the same job. It is only necessary that their zeroes should be correct. It is of no importance to us which one of such a group of functions is in use in the system; it operates in just the same way with each of them. We can call any two functions with exactly the same zeroes equivalent.

The need for discriminately closed subsets

An unexpected limitation arises on the sets S against which new elements are to be tested. We can understand this better if we look in two ways at the labelling of the first few elements. Let us first look at an oversimplified case in which it is possible to proceed by successively testing elements. The first element is labelled 1 because it is the least label. Further elements are tested against this until finally one appears which is not the same. This may be called 2, in which case the result of its testing will be (1,2). There is therefore a dcs of three elements against which to test the next element. When one appears which does not belong to this dcs it may be labelled 3 and as a result the testing of it against the three elements of the original dcs (a procedure that we argued could not take place without an infinite regress, but which we are assuming possible here for the sake of simplification) will give rise to (1,3), (2,3), (1,2,3). That is to say, there is now a larger dcs against which new elements are to be tested. This goes on. At every stage the set against which an element is to be tested is always a dcs.

Even when we take account of the fact that this oversimplified procedure cannot be followed, it still is necessary to consider only dcs's in order to ensure an unambiguous labelling. The reason for this lies in the way in which at every stage the process may introduce a new element or may be in the midst of determining whether a putatively new element is indeed new. This result means that the process consists of a sequence of generations and discriminations. This can be explained by using some particular process, such as that using the multiplication function mentioned above. It is not necessary to go into all the details of note 2 so long as the function defined there is taken for granted. The first element is labelled 1 as usual and the function for the set $S_1 = [1]$ is then $S_1(x) = x + 1$. When a new element arises it can be labelled

2 and the signal that it is indeed a new element is then labelled $1 + 2 = (1, 2)$. The set $[1, 2, (1, 2)]$ is then a dcs, and it gives rise to a function $S_3(x)$ where the suffix refers to the number of elements. In fact, $S_3(x) = (x+1)(x+2)(x+(1,2))$ but we need not enter into such algebraic details here. When a new element arises it can be labelled 3 and the signal that it is indeed a new element can be calculated by the methods of note 2 to be (1,2,3,4). This exhibits a first defect of the method; at this stage only the elements of the discriminate closure of [1,2,3] are relevant to the physics, but the multiplication function has gone outside this set to introduce 4. This feature, which we refer to as not being "compact", is undesirable because 4 is no longer available as a new label. But it is a feature of the multiplication function in particular and can be avoided by using other functions.

The set $[1, 2, (1, 2), 3, (1, 2, 3, 4)]$ is not a dcs, but none the less it gives rise to a function $S_5(x)$. When a new element arises it will be labelled with the least unused label, which is (1,3), and the signal that it is indeed a new element is then found to be 2. This exhibits a second defect of the multiplication function; for this signal is a label of a certain process — that of checking whether the element is new, and 2 was a label of a second element found earlier. There seems no reason why these two elements should be equivalent. This can be called being "re-entrant", and it is again a defect of the multiplication function which can be avoided by using other functions.

The third difficulty is the one which really concerns us because it is a common feature of any function, not only of the multiplication one. It has therefore to be dealt with by a more fundamental change. It is caused by the fact that the set against which the next element is to be tested is not a dcs. What is the consequence of this? When another element is introduced into the system its labelling is not determined unambiguously because it may be that in the course of various discriminations the label has been generated already. In that case the new element is to be labelled 4. But if not, then the least label, which should therefore be given to it, is (1,3). In that case, however, if in later discriminations the element (1,3) is produced, it may or may not be the case that this is the same element as that labelled (1,3) earlier. The nature of the trouble is now clear. Once certain labels have appeared, all would be well if one could consider the discriminately closed subset generated by all elements in play up to now and ask, not whether a "new" element has been generated already, but whether it belongs to this discriminately closed subset. We shall need to discuss later the extent to which this restriction is physically possible. Problems arise because we cannot take the whole set to be given, since it is in the process of growing. It will be necessary to have some conventional rule: rules like this, in which a number of tests are made and it is decided to accept a particular answer if it has been obtained a certain number of times will correspond to some physical parameter in the physical process. An example of this kind might be the restriction of the maximum energy in some scattering process to some value. For the present it is clear that the process must operate in such a way that a set is not a candidate to have a characteristic function until it has become discriminately closed. We could say that S will not be available until the discriminately closed subset has been "filled up". This is necessary to prevent ambiguity in labelling; we shall show that it is also sufficient by setting up such a scheme of precise characteristic functions.

Construction of characteristic functions

We consider, then, any dcs S. The avoidance of ambiguity has led to a greater importance being assigned to dcss in the development of the system. It is therefore useful to introduce some shorthand notations. Let U be any set of elements, not in general discriminatorily closed. Then $S = D(U)$ will denote the discriminate closure of U (that is, the set which results from taking U and adjoining to it the results of all possible discriminations between differing elements of U, and between elements so produced). $D(U)$ will have $r^* = 2r - 1$ elements: r is said to be the number of independent elements of U. If U is specified by its elements $U = [u_1, u_2, \ldots, u_r]$, it is convenient to drop the brackets and to write, instead of

$$S = D([u_1, u_2, \ldots, u_r]), \qquad S = D(u_1, u_2, \ldots, u_r) \,.$$

Similarly, if another element u is adjoined to U and the discriminate closure $D([u] \cup S)$ is formed, it is convenient to write $D(u, S)$ for it. Finally, if in a special case, U is the set $[1, 2, 3, \ldots, r]$, then one writes D_r for $D(U)$.

Our guide in finding a characteristic function for S is the method used (Note 1) to find one in the case where S has one element only. That is to say that the Principle of Choice will be used to construct a process amenable to analysis and equivalent to any other. It will be recalled that in that case the process was restricted by the three conditions

(i) $U(u) = 0$,
(ii) If $U(v) = 0$, then $V(u) = 0$,
(iii) If $U(v) = 0$ and $U(w) = 0$, then $V(w) = 0$.

It was shown then that it is always possible to find amongst the processes one which satisfies (i) and also the stronger forms of (ii), (iii):

(ii)' $U(v) = V(u)$,
(iii)' If $U(v) = U(w)$, then $v = w$.

Of the three conditions, (i) is simply part of the definition; and the corresponding condition for a general dcs, S, is
   (i) $S(u) = 0$ if and only if u is in S.

Condition (ii) has no simple analogue here because of the lack of symmetry between a set S and an element u, but the analogue of (iii) has the form:
   (ii) If $S(v) = 0$ and $S(w) = 0$ then either $v = w$ or $S(v+w) = 0$,
but not both. This analogue is simply a statement that S is a dcs. To avoid the repetition of the "but not both" phrase it is preferable to write it in the logically equivalent form:
   (iii) If $S(v) = 0$ and $S(w) = 0$ and $v \neq w$, then $S(v+w) = 0$.
We now follow the guidance of the simpler case to show that there is one process amongst those operating (though in contrast with the simpler case, this will not be unique) which satisfies (i) and the stronger form of (iii):
   (iii)' If $S(v) = S(w)$ and $v \neq w$, then $S(v+w) = 0$.
The method of doing this is by a recursive construction like the one used before. Of course, since the value of $S(u)$ has to be defined for any u, some sort of recursive construction is needed to provide for new u's as they are introduced. The recursive construction consists essentially in choosing for $S(u)$ as u proceeds upwards the least element in play up to that point which satisfies each of the constraints including (i) and (iii)'. (The phrase "in play up to that point" needs some qualification because of the possibility of arbitrary discriminations. The set of elements in play up to a certain point means the discriminate closure of the set of all elements which have been mentioned in the evolution of the constructions up to that point.)

The algorithm

We give the recursive construction in the form of an algorithm which spells out in detail the choosing of the least element in play which is consistent with (i) and (iii)'.

1. If u is in S, define $S(u) = 0$. This then defines S on the dcs S.
2. If v is the least element outside S, define $S(v) = u_1$ where u_1 is the least element of S.

3. For every u in S it is also the case that $S(v+u) = u_1$ (because any two such elements discriminate to $u+u'$, that is to an element of S so that (iii)' is satisfied). This defines S on the dcs $S_1 = D(v, S)$.
4. If v_1 is the least element outside S_1, define in the same way

$$S(v_1) = u_2, \qquad S(v_1 + u) = u_2$$

for any element u of S, where u_2 is the least element of S_1 apart from u_1, that is of $S_1 - \{u_1\}$.
5. It remains to define the value for $S(v_1 + w)$ where w is any element of S_1 but not in S. The restriction (iii)' implies that this cannot be u_2 or u_1 so it is to be taken as the least element apart from them, which is in fact $u_1 + u_2$. This now defines S on the dcs

$$S_2 = D(v_1, S_1),$$

and the possible non-zero values of S form the dcs $D(u_1, u_2)$.
6. In exactly the same way, if v_2 is the least element outside S_2 then $S(v_2) = u_3$, $S(v+u) = u_3$ for any u in S, where u_3 is the least element of $S_2 - D(u_1, u_2)$. Also $S(v_2 + v_1) = u_2 + u_3 = S(v_2 + v_1 + u)$. This now defines S on the dcs $S_3 = D(v_2, S_2)$, and the possible values of S form the dcs $D(u_1, u_2, u_3)$. It should be noticed that the result of choosing the least element satisfying the two constraints, in the last case, could be expressed more succinctly as:

$$S(v_2) = u_3, \qquad S(v_2 + x) = u_3 + S(x) \text{ for any } x \text{ in } S_2 \ .$$

This allows the general step of the recursion to be expressed in the form: If S has been defined for the dcs S_r, and v_r is the least element outside S_r, then, if u_{r+1} is the least element of $S_r - D(u_1, u_2, \ldots, u_r)$

$$S(v_r) = u_{r+1}, \qquad S(v_r + x) = u_{r+1} + S(x) \text{ for any element } x \text{ in } S_r \ .$$

This then defines S for $S_{r+1} = D(v_r, S_r)$.

In order to understand this construction fully it is instructive to work out some examples. Take S first to be the dcs $[1, 23, 123]$. The first step gives $S(1) = S(23) = S(123) = 0$. Next, $S(2) = 1$ and so $S(12) = S(3) = S(13) = 1$. The third step gives $S(4) = 23$, and so we have the table:

$u =$	14	24	124	34	134	234	1234
$S(u) =$	23	123	123	123	123	23	23

Next, $S(5) = 2$ and so we have the various values 1, 2, 12, 3, 13, 23, 123 given to elements up to 12345; this means that $S(6) = 4$, and it is clear that for any basic $r \geq 3$,
$$S(r+1) = S(r) + 1 \; .$$

The construction therefore leads to a form of $S(u)$ which is defined for all u, but with an infinite tail of values which carries no information at all since the values are determined as a recursive function. The only important part of S is the initial one.

$u =$	1	2	12	3	13	23	123
$S(u) =$	0	1	1	1	1	0	0

Secondly, take the case where $S = [3, 4, (3, 4)]$. The first two steps of the algorithm then give $S(1) = 3$, $S(2) = 1$. Notice how the algorithm ensures its validity by requiring the least element of $S_r - D(u_1 \ldots u_r)$. Otherwise one might have selected $S(1) = 2$, $S(2) = 1$, and this will not do because then the algorithm will give $S((1,2)) = (1,2)$. The rest of the calculation is straightforward and gives

$u =$	1	2	$(1,2)$	3	$(1,3)$	$(2,3)$	$(1,2,3)$	4	$(1,4)$	$(2,4)\ldots$
$S(u) =$	3	1	$(1,3)$	0	3	1	$(1,3)$	0	3	$1\ldots$

It will be clear from these examples that the process is indeed algorithmic, because there are always more values to choose from. The question of how much one is forced to go up to new values is one of importance and we shall come to it later.

The algorithm above serves also as a proof of the theorem:

Theorem 6.5. For any dcs there is a characteristic function satisfying the conditions (i), (iii)$'$ and such that, if u, v are any two different elements, then

$$S(u + v) = S(u) + S(v) \; .$$

Linear characteristic functions

In order to exhibit the algebraic structure in more conventional form, as we did in the case of S having one element, it is useful to augment the definition of S by defining
$$S(0) = 0 \; .$$

This additional definition is always permissible, because 0 is not an element and so will not present itself as an argument for S in the operation of the system. With this augmentation of S it is possible to drop the restriction on u, v being different elements in the statement of the theorem since, if $u = v$, $S(u)+S(v) = 0$ and $u+v = 0$. In future, we shall understand by a characteristic function one augmented in this way. The theorem then equally states

Theorem 6.6. For any dcs S there is a linear characteristic function satisfying the conditions (i), (iii)$'$.

In order to operate more conveniently with these characteristic functions, we go back to the first detailed argument above. It is clear in this case that the infinite "tail" of values beyond that for (1,2,3) need not be written down, since it contains no additional information. The reason why it does not is that, given the initial part (up to the value for (1,2,3)), the rest can be written down as the values of a recursive function. To see this in general, observe that it is only necessary to know the zeroes of S. If n is the maximum order of elements of S then no zero can occur in a position of higher order. Moreover, in giving this initial part, the whole table given in the example need not be set out. It would be sufficient in the examples given, to specify S for the basic elements $1, 2, 3$ because the linearity allows, for example, the calculation of $S((1,2)) = S(1) + S(2) = 0 + 1 = 1$. These two considerations suggest that we define a new type of entity, which we call an *array* in the following way:

Let there be k rows $r_1, r_2, r_3, \ldots, r_k$ none of which is of order greater than k. Then an *array* is defined as

$$R = [r_1; r_2; r_3; \ldots ; r_k] .$$

Such an array represents a linear operator by the rules

(i) if s is a member of L, $R(s) = r_s$
(ii) if u is not a member of L, $R(u)$ is got by expressing u in terms of basic elements and using linearity.

In the vector space notation, where the dimension of the r's is arbitrarily fixed at k, an array just behaves like a square matrix. So arrays are simply the necessary generalization we have to make to deal with rows (in our sense of the word) rather than with bit-strings of fixed length. The augmented characteristic functions in the theorem are not arrays since the order of the elements is not restricted. But the truncated form, which contains all the information, can be so represented. This was exhibited in the examples.

We now summarize the position. The process picture commits the system to restricting its testing to dcss. Whatever description is used for these dcss, it is equivalent to linear (augmented) characteristic functions which can be represented by their truncated form as arrays. We have therefore clarified the introduction of matrices (finite dimensional linear operators), which, like discrimination, looked a bit ad hoc in Chapter 5.

There is one case which needs a little more attention: when the dcs has the form of the discriminate closure of the first n elements of L, D $(1, 2, 3 \ldots, n)$. Take the example $n = 3$. The array of order 3 which is the truncated form of the function for this set is then the zero array $(0; 0; 0)$. But we showed above that the set of characteristic functions for a discrimination system was itself a discrimination system, and in using this result in the Parker-Rhodes construction it would be convenient to use the truncated form of the functions. To do so in this case, however, would run into conflict with our earlier convention of excluding zero elements. We can avoid this conflict by recalling that the zero array here is the minimum representation of the information, but could also be replaced by the array $(0; 0; 0; 1)$. This is the course we shall continue with for the present, though historically it is not what was done. Instead the issue was dodged with a clever device of Parker-Rhodes. Instead of using a truncated array R to characterise a set by the rule

$$R(x) = 0 \text{ if and only if } x \text{ is in } S \ ,$$

he considers (as we explained in Chapter 5) a matrix A which characterises S by $A(x) = x$ if and only if x is in S.

Since arrays are matrices and $R = S + I$ where I is the identity matrix, the two descriptions are completely equivalent. We shall go over to the Parker-Rhodes version in Chapter 7, but for the present we remain with ours to avoid a major break in the argument.

The Parker-Rhodes bounds

We now turn to a more detailed discussion of level change. We pointed out that the process was not constrained to change level, but the general ergodic principle which we assumed means that the possibility of level-change will eventually be realised. Level-change can be viewed as an increase in the self-organization of the system, since ascent to a higher level means that whole sets of elements are organized to appear as single elements. The Parker-Rhodes construction described in Chapter 5 is therefore a determination of the maximum degree of self-organization possible for the process, and the numbers 3, 10, 137, 10^{38} arising in it are bounds on this self-organization. The construction translates into our terms. It all depends on the result just mentioned, that the set of characteristic functions itself forms a discrimination system.

This result is easily extended to show that it holds also for the set of truncated arrays. The Parker-Rhodes construction begins with 2 elements which give rise to 3 dcss, as explained in Chapter 5. The 3 arrays for these constitute the next level — giving rise in their turn to 7 dcss. Parker-Rhodes assumed that it is possible to choose the 7 functions for these sets so that at the next level there would be 127 dcss. This is in fact true; we should express the situation in terms which fit better in our system by saying that if the sets of 7 are chosen randomly, some will have the property of giving a higher level with 127 dcss. (In fact, a majority will have this property.) The 127 arrays will give rise to too many dcss for their functions to be incorporated into the vector-space picture of Parker-Rhodes. That is because the original elements are 1,2 in our notation — that is, 2-dimensional vectors. The linear arrays for them are 4-dimensional (although we are concerned only with a 3-dimensional subspace), so that the 7 arrays at the next level are $4^2 = 16$-dimensional. The 127 at the next level are $256^2 = 65536$-dimensional, although there is need for 10^{38} of them, and the construction stops.

Little is changed by our new apparatus except that more needs to be said about the bounding numbers. It might be thought that at the level with 7 elements the effective dimension is 3 although Parker-Rhodes works with 4×4 matrices. This seems to imply that the dimensionality is to some extent arbitrary, which would mean that the requirement to find operators at the top level from a 65536-dimensional space is also arbitrary. There is indeed an arbitrary aspect about Parker-Rhodes' description of his construction, but it is an arbitrariness, like that over the introduction of discrimination in place of eigenvectors of matrices, which our exposition is able to clarify. Consider any set of r elements $[1, 2, 3 \ldots r]$ which generate a dcs of r^* members. Arbitrary discriminations between these members will always give one of them, or zero (i.e. will give one of 2^r cases). To specify a member is therefore to give r bits of information: for shortness let us say each element carries r bits. Of course some sets may carry less; for example, for $[1, 2, 12, 3]$ $r = 4$, but the dcs generated has only $3^* = 7$ elements, so every element carries 3 bits in this case.

Return now to the case of r elements, each carrying r bits. There are r^* characteristic functions, one for each of the dcss which are subsets of the whole set. Any such function is specified by listing the r elements (out of a possible r^*) into which $1, 2, \ldots, r$ respectively are carried. For each of $1, 2, \ldots$ the listing can take the form: Is 1 present in the answer? and so on. These r bits arise for each element, and so the functions carry r^2 bits. One can start off a hierarchy construction provided that $r^* < r^2$, which limits r to $2, 3, 4$. So much is straightforward. Parker-Rhodes now appeals to square matrices to avoid the following argument: At the next stage there will be r^* elements each carrying r^2 bits, and, for this to be possible, $r^2 > r^*$. In order to carry on as at the previous stage, one must regard the r^* elements as a subset of r^2 independent ones, each carrying r^2 bits, so as to allow the same construction again. The next step is then possible so long as $(r^*)^* < r^4$, and this limits r to 2. That the effective dimension for doing the algebra is r^* is true enough (and this fact will be utilized for doing the detailed calculations frequently) but the rest of the argument about arbitrariness ignores the need for information-carrying which we have just explained. This need again produces the bounding numbers 4, 16, 256, 65536 as for Parker-Rhodes. The same numerical bounds therefore result. It would have been surprising if this had not been the case, since the same bounding numbers arose already with the multiplication functions, indicating their universality.

Proof of the Parker-Rhodes theorem

For a time during the early discussions of the constructions the proof eluded us that it was possible to find the required 7, 127, arrays which would fulfil all the requirements including that of giving rise to the full number of dcss at the next level (i.e. of being linearly independent). To get 7 out of 16 does not seem surprising, but 127 out of 256 seems to be living slightly dangerously. Noyes showed that there were 127 matrices by constructing them, but it was too tedious to show that they were linearly independent. Kilmister gave a quite different proof that the set of 127 linearly independent matrices existed, but it was only an existence proof, and did not give any indication of what they were. The key question is how big the dimension of the vector space has to be to accomodate all the constraints. Suppose that there are r elements and so r^* dcss; then it is obvious that there is no hope for linear independence if the order of the arrays is n where n^2 is less than r^* since n^2 is the dimension of the vector space of the arrays. It is rather surprising that this is also enough.

Theorem 6.7. Given r elements, so r^* dcss: their arrays have truncated forms which are arrays of order n if and only if $r^* < n^2$.

The following proof is not the most elegant, but it is instructive because it shows why the result is true. (Since the particular application to $r = 7$, $r^* = 127$ gives $n = 12$ as enough, rather than $n = 16$, there is some need to see why this can be true.)

Stage 1. Begin with the case $r = 2$ and use the algorithm above to construct the arrays for the dcss [1], [2], [1,2,12] which we write for shortness as 1,2,12 respectively. The result can be tabulated by giving the value of the array for 1,2 respectively:

S	1	2
1	0	1
2	2	0
12	0	0

If the first term in the infinite tail is included, the table becomes

S	1	2	3
1	0	1	2
2	2	0	1
(1,2)	0	0	1

and the three arrays are obviously linearly independent. This situation is not typical of the higher levels. Because of the linear independence, we may denote the three arrays by 1, 2, 3 and proceed to the next level, $r = 3$.

Stage 2. A first shot gives the following table.

S	1	2	3
1	0	1	2
2	2	0	1
12	0	0	1
3	3	1	0
13	0	1	0
23	2	0	0
123	0	0	0

Each line of this table is good by itself, but the signals are not linearly independent as can be seen by adding together the ones for 2, 12, 23. If the beginning of the infinite tail is included, the table becomes:

S	1	2	3	4
1	0	1	2	3
2	2	0	1	3
12	0	0	1	2
3	3	1	0	2
13	0	1	0	2
23	2	0	0	1
123	0	0	0	1

The resultant arrays form a linearly independent set, as can be seen at once because in any sum of them which is to come to zero the first one cannot be included because of the 2 in the third column; then the second cannot be because of the 3 in column 4, and so on. It is however untypical for the inclusion of the tail to put matters right in this way, and so we need a general procedure for dealing with this situation. This is to go back to the last array which causes trouble and try to modify it so as to cure the trouble. Here it is the array for 23. The 2 needs to be changed, but of the six other values it might have, 1 is excluded by diversity, and so is (1,2) which we shorten as 12, and so is 13. 3 is excluded by linear independence, and so is 23, and 123 is excluded by diversity. Accordingly one moves back to the next row contributing to the trouble, which is that for 2. Here it is easy to see that 2 cannot be replaced by 1, 12, or 3, but 13 will serve to meet all the requirements. A similar piece of work serves to do the case $r = 4$, so showing that the arrays for the $4^* = 15$ dcss can be chosen in 16 dimensions, which is surprising. We omit the details here; they are to be found in note 3 at the end of the chapter. At the next stage, however, the $5^* = 31$ arrays need $6^2 = 36$ dimensions, and it is much easier to satisfy the linear independence since the table to be filled in has 6 columns instead of 5. The extra column can be used to help over the linear independence. At each later stage in the theorem these extra columns are present in increasing numbers. For example, for $r = 7$, which is the next case needed in the hierarchy construction there are $7^* = 127$ dcss and so the number of columns is 12, giving great freedom to rectify any troubles.

This chapter has served to show how the Parker-Rhodes construction, or something equivalent to it, is implied by the process view which we are proposing. In the course of doing this, it has become apparent that neither discrimination nor the use of matrices to characterize sets are arbitrary devices, but are intrinsic parts of the system. The theorem that the Parker-Rhodes hierarchy actually exists has been proved. In succeeding chapters we shall show how these ideas can be applied in physical problems.

Note 1

Any equivalence relation, $=$, satisfies the conditions:

(i) $s = s$ for any s,
(ii) If $s = t$, then $t = s$.
(iii) If $s = t$ and $t = u$ then $s = u$.

In terms of the testing, one can write these (again not distinguishing between the set U and the process of determining its members) as:

(i) $U(u) = z$ for any u.
(ii) If $U(x) = z$ then $X(u) = z$.
(iii) If $U(v) = z$ and $U(x) = z$, then $x = v$.

(The last one of these has been slightly changed, for later use, by using the second.) There is no more to be said about the first of these conditions; it is simply implied by the definition of the set Z. The other two allow for some change to make a system which is more convenient to analyse and yet works like the original system.

Consider first the condition (ii). The operator U is not restricted by (i) and (ii) except to the extent of giving the value z only for argument u. Similar remarks apply to X. It is obviously possible, starting from two particular operators U, X to derive a new one by the rule

$$U'(x) = X(u), \text{ if } u > x, \qquad U'(x) = U(x), \text{ if } u < x.$$

Interchanging u, x gives

$$X'(u) = U(x), \text{ if } x > u, \qquad X'(u) = X(u), \text{ if } x < u.$$

Here note that u, x are any pair of labels. Now $U'(x) = z$ only in the case when one (and therefore both) of $U(x), X(u)$ takes the value z. That is to say, the new operator U' is equivalent to U and the system proceeds in exactly the same way whichever of them is used. It is therefore permissible, when analysing the process, to assume that the operators are in fact U', X'. Dropping the primes, then, it is permissible to add the extra condition, which can be seen as a strengthened form of (ii) above:

(ii)' $U(x) = X(u)$ for all u, x.

We shall now show that it is possible to strengthen the condition (iii) in a similar way, to (iii)′ If $U(v) = U(x)$, then $v = x$. At the same time we shall show that a further restriction is possible. This one does not arise from the properties of the equality relation, but is simply a convenience in the analysis. Both of these restrictions still leave one with operators which are equivalent to the original ones, so that the process works in just the same way with them. The further restriction is the

Assumption. For any set S, $S(x) \neq x$.

The usefulness of the extra assumption is that it provides a clear indication that the process S has actually operated, since x is changed. The general idea behind the strengthening of (iii) is that the condition (iii)′ just states that all the values of U are different. This can be ensured only by some sort of recursive definition of U (for how else could one deal with all the values?) and the proof consists in spelling out the necessary recursive construction. We refer to it as 'Conway's trick' because it is equivalent to a corresponding definition in a different context and in a different notation in his *On Numbers and Games*. In order to spell out this recursion, let OP be the set of operators, like U, which satisfy (i), (ii)′ and assumption 5. Then the recursive definition of a new function U^* which will be equivalent to the old ones is:

$$U^*(x) = \min[U(x) : U \text{ in } OP \text{ and } U(x') \neq U(x) \quad \text{for any } x' < x] \, .$$

Here it is understood that the minimalization is over different equivalent U's. It is obvious that, on the one hand, U^* will satisfy (iii)′; a formal proof being just a recursive one. On the other hand, since the minimum is for some U in OP, the newly defined U^* will be in OP. So we drop the asterisk on U and simply suppose that we are dealing with such a U.

We shall show shortly that the restrictions by which we have tied down the various equivalent operators are now sufficient to specify the operators completely, in terms of the ordering of the labels which we postulated earlier but have not yet specified. This means that we have been able to satisfy the mathematician who saw the operators U and so on as functional symbols. It is convenient to adopt a different notation in anticipation of this. If, as usual, u is the unique element of U, we replace $U(x)$ by $[u, x]$. Then the three restrictions that arise from strengthening the idea of an equality relation become:

(i) $[u, u] = z$,
(ii) $[u, x] = [x, u]$,
(iii) if $[u, x] = [u, y]$, then $x = y$.

This has all been carried out in terms of any specified ordering of the strings in $W(L)$. The only restriction that has been placed on the ordering is that the non-label symbol, z, comes at the beginning. What has to be done next is to specify the ordering, and this will automatically have the effect of saying how the symbols in L are to be used.

It is necessary for this purpose to adopt some conventions about the symbol z. It is clear that the actual process under consideration can never throw up expressions like
$$[z,u], \quad [u,z], \quad [z,z],$$
but these forms may come up in the course of the working, in the reduction of labels to simpler forms. It cannot affect the physical process in any way if we give what values we like to these expressions. Because of condition (ii) it is possible and convenient to define the ordering only for strings in L of the form
$$r = (r_1, r_2, \ldots, r_k) \text{ with } r_1 \leq r_2 \leq r_3 \leq \ldots \leq r_k.$$
Also because of condition (i) the possibility of equality can be ignored so that $r_1 < r_2 < \ldots < r_k$. One can then order these words by reverse lexicographical ordering:
$$1, 2, (1,2), 3, (1,3), (2,3), (1,2,3), 4, \ldots.$$
Carrying out this construction for the operation $[u, x]$ recursively, the first two labels, 1, 2 need no comment. Then [1,2] yields the string (1,2). Consider next the value of ([1,[1,2]) which is [1,(1,2)]. From the conditions on $[u, v]$ this cannot have the values 1, (1,2), z, so the least value it can have is 2. This is the value which the rules give it. Similarly [2, (1,2)] will have the value 1. Thus the three elements 1, 2, (1,2) have a structure isomorphic to the quadratic group, except that the identity (zero) element has been omitted. The omitted element is z and so we shall write it as 0 in future and call the testing operation $[u, x]$ *discrimination*, and denote it by $u + x$. The quadratic group is the first non-trivial example of a *discriminately closed subset*. Continuing the recursive process in the new notation, the next element is 3 and the values of $1+3$, $2+3$ come out at once. That of $(1,2) + 3$ cannot be $2, 1, 0, (1,2), 3, (1,3), (2,3)$ so must be taken as the next one, $(1,2,3)$. It is clear how things are going and a few minutes consideration will show that $u + x$ as defined here is the same as $u + x$ (differently defined) in the text.

Note 2

The simplest way of defining a product (without divisors of zero) is, following Conway (loc. cit.) to define xy as the least number not forbidden by the commutative and associative rules (and the requirement that it be new) and also take account of distributivity over addition. Thus taking $2 \times 2 = a$, and going over into the abbreviated notation of 12 for (1,2) gives the table.

	1	2	12
2		a	$2a$
12		$2a$	$1a$

This shows that a cannot be 1 or 2, but $a = 12$ is possible. The table for the set $D(1, 2)$ is now closed under multiplication and the structure is that of the cyclic group of order 3, C_3 and is therefore associative, as it should be. When a new element, 3, is introduced we assume that $2 \times 3 = a$ and that $3 \times 3 = b$. We can then construct the partial table below. This table shows at once that a cannot be 1, 2, 12, 3, 13, 23, or 123, so a must be taken as 4. Then b cannot be 3, 1, 234, 4, 124, 34, 13, 12, or 2, leaving the possibility that $b = 23$. That this is indeed possible is shown by the fact that the resultant table of 15 elements is isomorphic to the group $C_3 \times C_5$ (for 4 is an element of order 5 which therefore generates a C_5, and the product of this with the C_3 above generates the required table).

	1	2	12	3	13	23	123
2		12	1	a	$2a$	$12a$	$1a$
12		1	2				
3				b	$3b$	ab	$3ab$
13							
23						$12b$	$13b$
123							

To generalize this it should be noted that two kinds of new elements enter: those (like 3) which are introduced to extend the table which has become closed, and those (like 4) which arise internally from the operation of the extension. Elements of the first type will be called basic, and the next basic element is evidently 5. A similar argument will show the need to have 3 new elements of the second type to account for 2×5, 3×5, and 4×5, and also that 5×2 may be taken as 45. Calling these 3 new elements 6, 7, and 8 means that the next basic element is 9. In its turn it will need internal new elements for $n \times 9$ ($n = 2, 3, \ldots, 8$) so these will be $10, 11, \ldots, 16$, and the next basic element will be 17. The basic elements form the set

$$F = [2, 3, 5, 9, 17, \ldots] = [f_0, f_1, f_2, \ldots]$$

where
$$f_r = 2^r + 1 = r^* + 2, \text{ or more usefully, } f_r = 2f_{r-1} - 1 .$$

These results suggest strongly an alternative form of the product, also given by Conway, which has the advantage of avoiding the explicit use of the "least number such that": If s is in F and $r < s$, with r, s in L, then $r \times s = (r+s-1)$, $s^2 = (s-1, s)$.

All other products are to be found from these rules and by the use of commutativity, associativity and distributivity over addition. It may be useful to have an example to show how the calculations are carried out. Take

$$234 \times 34 = (2 + 3 + 4)(3 + 4) = 2 \times 3 + 2 \times 4 + 3^2 + 4^2 .$$

Next one has to use the fact that $4 = 2 \times 3$ (and the nature of the proof below of the correctness of this definition of a product is to show that such substitutions can always be found). The product becomes:

$$4 + 2^2 \times 3 + 2 + 3 + 2^2 \times 3^2 = 4 + (1+2) \times 3 + 2 + 3 + (1+2) \times (2+3)$$
$$= 4 + 3 + 4 + 2 + 3 + 2 + 1 + 2 + 3 + 4 = 1234 .$$

To justify this definition of a product we have to show three things. Firstly we must show it is complete (any product can be found). Then we must verify that the three laws hold for any product. Finally we must show that there are no divisors of zero (that a product never vanishes unless a factor in it vanishes). Begin by setting out the effect of the first part of the rules for elements of L:

	s			
$r=1$	2	3	5	9...
2		4	6	10...
3			7	11...
4			8	12...
5				13...
.				...
.				...

The table makes it clear that every element of L can be written uniquely as a product $r \times s$ with s in F and $r < s$. Moreover $r < r \times s$, so that the factorization can be repeated on the smaller number r. So, in a finite number of steps one has any element of L expressed as a product of members of F. This shows (with the corresponding rule for squares) that the definition is complete.

Next as to the three rules: they are imported by the use that is made of them in extending the definition to all products, and a moment's consideration is enough to show that there is no question about the commutative or distributive laws. As to the associative law, that could only fail if if there were a product $(r \times s) \times t$ with r, s, t all in F which was different from $r \times (s \times t)$. But how can the latter product be found? Unless $s \times t$ is in F it must be analysed further as a product and so $(r \times s) \times t$ is used; so a counterexample to the associative law can arise only if $s \times t$ is in F. A glance at the table shows that this occurs only if $s = 1$, and in that case the associative law holds trivially.

Finally, the proof that there are no divisors of zero is a simple induction on the number of elements involved.

Note 3

Stage 3. $r = 4$

In the following table, to assist in making the procedure clear, the original form of each array is given, and the amended values given in brackets:

	1	2	3	4
1, 2, 3, 4	0	0	0	0
1, 2, 3	0	0	0	1
1, 2, 4	0	0	1	0
1, 3, 4	0	1	0	0
2, 3, 4	2	0	0	0
1, 2	0	0	1	2
1, 3	0	1(3)	0	2
2, 3	1(3)	0	1	1
1, 4	0	1	2	0
2, 4	1(3)	0	1	0
3, 4	3(4)	1	0	0
1	0	1	2	3
2	2	0	1(34)	3 P
3	3	1(24)	0	2
4	4	1(14)	2	0

The procedure is self-evident until the line marked P. Here the original form introduces no new basis elements and it is not possible to increase the first element to introduce one, since all have been used up in the first column. Similar considerations apply in the remaining two lines.

90   *Combinatorial Physics*

Stage 4. $r = 5$

	1	2	3	4	5	6
1,2,3,4,5	0	0	0	0	0	1
1,2,3,4	0	0	0	0	1	2
1,2,3,5	0	0	0	1	0	2
1,2,4,5	0	0	1	0	0	2
1,3,4,5	0	1	0	0	0	2
2,3,4,5	2	0	0	0	0	1
1,2,3	0	0	0	1	2	3
1,2,4	0	0	12	1	2	3
1,3,4	0	13	0	0	2	3
2,3,4	2	0	0	0	1	3
1,2,5	0	0	1	2	0	3
1,3,5	0	24	0	2	0	3
2,3,5	23	0	0	1	0	3
1,4,5	0	5	2	0	0	3
2,4,5	123	0	1	0	0	3
3,4,5	4	1	0	0	0	3
1,2	0	0	1	2	3	4
1,3	0	26	0	2	3	4
2,3	25	0	0	1	3	4
1,4	0	1	124	0	3	4
2,4	6	0	1	0	3	4
3,4	3	1	0	0	2	4
1,5	0	1	2	3	0	4
2,5	2	0	15	3	0	4
3,5	3	1	0	5	0	4
4,5	4	1	6	0	0	3
1	0	1	2	3	4	5
2	2	0	1	346	4	5
3	3	1	0	245	4	5
4	4	1	2345	0	3	5
5	5	123	2	3	0	4

CHAPTER 7
Scattering and Coupling Constants

So far we have only discussed the numerical grid or frame provided for physical theory by the combinatorial hierarchy. In the next three chapters we go further and make contact with existing theory. In Chapter 9 we exhibit the use of the hierarchy model in the construction of space. In Chapter 8 we discuss the idea of the quantum number and the particle and give our alternative to the pattern provided by the standard model. In the present chapter we show how the grid can be regarded as part of a wider field of numbers capable of having a physical interpretation.

We do this by describing two important consequences of our combinatorial approach which are, unexpectedly, related. One of these is the development of the concept of time. The other is the calculation of the numerical value of the reciprocal of the fine-structure constant, $hc/2\pi e^2$. We derive as a close lower bound the value 137.03503 compared with the experimental value of 137.0359. It will become clear as the chapter proceeds that our thinking owes a great debt to two of our colleagues, Pierre Noyes and David McGoveran, although this does not necessarily mean their agreement with our present discussion. In Chapter 5 we explained the importance for us of Noyes' emphasis on the primacy of counting particle scatterings. He has also guided us through the thicket of particle physics. McGoveran has provided a complete theory of discrete quantities — the "ordering operator calculus" — which has been a source of inspiration to us. We have not used any of his detailed results because it would take us too far afield to set out his whole, very subtle, theory; and without doing that it would be dangerous to quote results out of context.

In the foregoing chapters we have encountered the coupling constants as pure numbers which are usually thought of as dimensionless ratios of the fundamental atomic and cosmical dimensional constants. We have preferred to call them 'scale-constants' because of their function in determining the relative strengths of the basic physical interactions and fields, and hence of providing the scaling framework of the world of physics. Indeed we have made the case for regarding the scale-constants as having a necessary role in their own right in the relations between the quantum world and the cosmos. It therefore comes as something of a surprise to find the most famous of them — the electromagnetic coupling constant — appearing as the parameter which specifies the fine splitting of the lines in the hydrogen spectrum. (It is known as the fine-structure constant for this reason.) There is a theory associated mainly with the name of Sommerfeld in which quite classical views of the nature of the electron (though in discrete orbits) are used to relate the quantities e, $h/2\pi$ and c. That theory, however, leaves the numerical value of the ratio $2\pi e^2/hc$ to be determined empirically. One is left wondering why the existence of this basic constant in the quantum theoretical picture should rest on this rather recondite calculation. The writers were perplexed by this thought for some years; up to about 1965. They knew that the basic interactions postulated by the combinatorial theory had to have an interpretation involving high-energy physics, but could not see how to interpret the scale constants in high-energy terms.

The new shape was provided by Noyes whose approach to quantum physics gave primacy to scattering processes and to the counting of events in scattering processes. Noyes has used the combinatorial model to press the claim that all knowledge in quantum physics consists in counting in scattering processes. Following this idea, the inevitability of the existence of coupling constants appears right at the outset. Suppose that there are entities of a certain number of different sorts. (We refer to particles, but would like to keep the argument at the abstract level.) Then, we must typify each sort of entity with its own transition probability. There is nothing else we can do, since by postulate all our knowledge cames from counting of statistical events. We can never draw aside the curtain to get privileged information about the entities, for that would require further scattering experiments of the kind we are trying to formulate, and lead to an infinite regress.

An important part of the original appeal of the combinatorial theory lay in the fact that it mapped a structure of scale constants which fitted pretty well the strange distribution of these constants in nature, with their very rapidly increasing gaps between them. This fitting retains its persuasive power though it is sometimes forgotten in the detail. Noyes and McGoveran[1] present other scale-constants in the context of the scattering idea. In particular, the fourth level scale-constant is interpreted as the coupling constant of the gravitational field $(hc/2\pi) \times (Gm_p^2)$, (or M_{Planck}/m_p^2), and a calculation is made of the coupling constant of the weak interaction. The latter is a refinement of an old argument of Bastin[2] which we quote: "The weak interactions have an experimental coupling constant 1.01×10^{-5}. This value is obtained by Feynman[3] by taking one of the possible particle masses, and is therefore subject to revision). The value we get from our theory is 1.53×10^{-5}. This value is obtained by considering the forbidden mapping $2^{127} \to (256)^2$, and arguing that if a constraint of some unknown nature were to be imposed on the elements of this level so as to reduce its effective multiplicity to $(256)^2$, this mapping would no longer be forbidden. Such a mapping would presumably correspond to unstable particles. The value $(256)^{-2}$ then would give the coupling."

To bring the combinatorial value of $(256)^{-2}$ into agreement with the value quoted by Feynman of 1.01×10^{-2} it is necessary (a) to note that Feynman's value is for the dimensionless combination $(2\pi G_F m_p^2)/hc$ and (b) that to compare this experimental value with the combinatoric one it is necessary to supply a conventional $\sqrt{2}$. The mass unit m_p^2 is required because G_F is a dimensional constant. That this is the mass unit in the combinatorial calculation is justified by the successful identification of $2\pi Gm_p^2/hc$ with $1/(2^{127}+136)$. The $\sqrt{2}$ requires a discussion of the difference between a Yukawa (3-particle) and a Fermi (4-particle) coupling which would take us too far afield. This simple type of argument in which corrections are deduced using statistical assumptions will appear in detailed form in Kilmister's calculations below.

There have been at least three attempts to calculate the value of the fine-structure constant, and we shall describe these now. The first two will detain us only briefly.

(a) EDDINGTON. Eddington was the first person to make a deliberate policy of deriving scale constants combinatorially. He derived the value 137 for $hc/2\pi e^2$ from a counting of numbers of degrees of freedom in an algebra. He never came to see that his method could not be sensible unless the scale constants were logically prior to less fundamental experimentally based numbers, and so he never faced the next problem which is how numbers describing spatial and dynamical relationships could be understood in his theory. The error in his calculation is 0.03%, but because of the way in which it was found there is no conceivable way of improving on it.

(b) WYLER.[4] Wyler finds the value

$$9/(8\pi^4) \cdot (\pi^5/2^4 5!)^{1/4} = 137.03608 \,,$$

which reduces the error to less than 0.00007%. He derives this by what he calls a "geometrical" calculation. By this he means the carrying out of some subtle analytical calculations of volumes in the space of the conformal group (the largest group under which the Maxwell equations are invariant). But the critical stage in his argument is as obscure as Eddington's (and is presented with quite an Eddingtonian flavour). The conformal group is isomorphic to $SO(5,2)$ and the group $SO(4,2)$ is also well known in relativistic particle kinematics: Wyler says, "The interpretation of the groups $SO(4,2)$ and $SO(5,2)$ leads to a model of the interaction between the Maxwell field and the field of elementary particles. This model is defined by the reduction of the representations of $SO(5,2)$ to $SO(4,2)$...." The objections to this are just the same as those given above for Eddington.

(c) MCGOVERAN. David McGoveran has carried out a thorough-going attempt to produce a discrete physics of a non-trivial kind, but starting from principles which, though consistent with those used in this book, are apparently different.

Because of the similarity of approach and the accuracy of his calculation of the fine-structure constant, we shall describe his work at greater length. In the course of the description we shall draw attention to the differences between his work and the present theory. These are to be read as no more than that, and not as criticisms of the internal consistency of his argument.

His argument has been put in several forms and the one we have chosen to describe at length is that in his contribution[5] to *Discrete Physics and Beyond*. An ordering operator is a finite directed graph. McGoveran prefers to think of the ordering operator as a finite computational object which generates a directed graph, but in extension the terms are almost the same. The difference is subtle but important. The full specification of the ordering operator requires the history of the generation of the directed graph. At points of the graph like (1), the ordering is partial, and A, B are called *indistinguishables*.

Fig. 1

At points like (2) the ordering is total. Possible walks of the graph include "jumps" between indistinguishables (involving "virtual arcs" which are undirected). We put inverted commas round 'jumps' to emphasize the indistinguishability of A and B. An unordered graph can be ordered in a variety of ways. (In McGoveran's language, "different ordering operators can act on the same graph".) A sub-graph represents a particular property ("combinatorial attribute", or simply "attribute") of the nodes. In terms of some combinatorial attribute a walk can be represented by a bit-string: if n is the number of nodes needed to traverse a sub-graph, one writes 0 if, after n nodes, the attribute has not been encountered (i.e. if these nodes do not form the subgraph) and 1 if it has. The bit-strings constructed in this way can be used to reconstruct the whole graph, so that the output of an ordering operator can be thought of as consisting of bit-strings. Hence the ordering operators evidently play somewhat the same role in McGoveran's theory that the underlying process plays in ours. One might say that the ordering operators are more explicit versions of (part of) the process.

McGoveran's starting point in a view which (like ours) is one of process, rests on the following basic ideas:

(a) The term "probability" is used to refer to counts of occurrences, so it is not usually divided by the total number of possibilities to normalize to 1. This is necessary for certain technical reasons connected with the difficulty of naming the total number of possibilities in a process theory.

(b) The existence of indistinguishables, as above, affects the notion of independence in probability. According to McGoveran there is a smooth (though discrete) spectrum of degrees of independence, ranging from mutual independence to complete dependence. He describes this spectrum as a multi-dimensional discrete, finite vector space, *a combinatorial attribute d-space*. A d-set of completely independent combinatorial attributes forms a d-basis of this space. Two or more completely independent attributes add as orthogonal vectors. All of this is to be understood taking account of the process point of view, so that two variates, x, y might become dependent when enough detail has been developed about the relationship between them, but until then they are independent.

(c) The Combinatorial Hierarchy, which is taken as one particular example of McGoveran's mathematical structures, gives 1/137 as the first-order approximation to the fine-structure constant.

(d) A "conceptual model of the hydrogen atom" is needed, and provided by McGoveran's argument, to give the second order correction.

(e) There would, in principle, be higher order corrections.

(f) In the representation by bit-strings, the basic notion in the argument is that of a *combinatorial event*, which is represented by a bit-string *pattern*.

The principal difference from our approach lies in (c) and (d). We shall argue that the derivation of a scale-constant having the value 1/137 is enough (given our position that the scale-constants are logically prior) to identify it as the fine-structure constant and that therefore we can use this identification when we turn to calculating the constant more carefully by seeing exactly how it arises in the theoretical system. McGoveran (d) differs from us in that he identifies the constant from the part it plays in a particular problem about the spectrum of the hydrogen atom rather than from its more general significance.

More should be said about the part of McGoveran's calculation which uses the combinatorics of the matrix mapping space in the hierarchy construction (apart from the problematical identification of frequencies with the Bohr atom). Our approach has been to import ideas of physical space and time only so far as we have built them up, and never making a once-for-all identification. However we are far short of a whole picture because we can as yet give no sense to a space in which different objects can have an independent existence. This boundary to our conceptual scheme will have to be pushed back by exploiting the recursive nature of the hierarchy in such a way that the steps through it are *recorded* as distinct from just happening. Our older vision of the hierarchy saw the matrix mapping space as furnishing apparatus for doing this. Even though this view has to be modified because at level 3 a space of smaller dimensionality will suffice for the construction it would be surprising if one had to discard the whole potential of it. McGoveran's calculation makes use of the matrix mapping space without explaining its relation to physical space which, from our point of view is jumping the gun. One hopes that his calculation of the fine-structure correction may be correct as well as Kilmister's — partly because that would be a strong indication of the form of the spatial use of the matrix mappings and their combinatorial properties.

If a combinatorial event recurs "in a regular way" it is called *periodic* and the number of bits generated between occurrences is the *cyclicity*. These ideas are now applied to the conceptual model of the hydrogen atom. It is a little difficult to tease out what this 'conceptual model' is, but it seems to refer back to the Old Quantum Theory of the hydrogen atom in terms of the two quantum numbers j, s (a treatment which comes up again naturally in the wave-mechanical treatment). Naturally the Bohr–Sommerfeld treatment is not taken over lock, stock and barrel, but from it are borrowed two important ideas:

(a) There is a two-fold cyclicity, corresponding to what were called elliptical orbits in the Old Quantum Theory;
(b) There is a basic probability 1/137 corresponding to what was put in as a Coulomb interaction in The Old Quantum Theory.

Accordingly, McGoveran considers a system exhibiting two distinct cyclicities which he calls s and j. Since s, j must be different one may take $s > j$. Now whatever the relative values of j, s, if they are truly distinguishable then they will not in general have any coincidences. McGoveran's argument is a little obscure at this point, so we try to elucidate it by some examples. The case $j = 2$, $s = 5$ is clearly ruled out by their being not distinguishable unless the situation sketched in Fig. 2 is modified by starting s off so as not to coincide with a j (see Fig. 3). In this case there is an s point (say s_0) which is nearest to a j. But it is not at all clear that this will always be the case. For consider Fig. 4, which is actually calculated by taking $j = 2$, $s = 1 + \sqrt{3}$ and starting off s from $s = 1 + \sqrt{3}$. It does not matter at all that McGoveran's theory is a discrete one, from which $\sqrt{3}$ has been excluded, since these figures could equally have been calculated by using suitable rational approximations, or indeed by means of a recursive function.

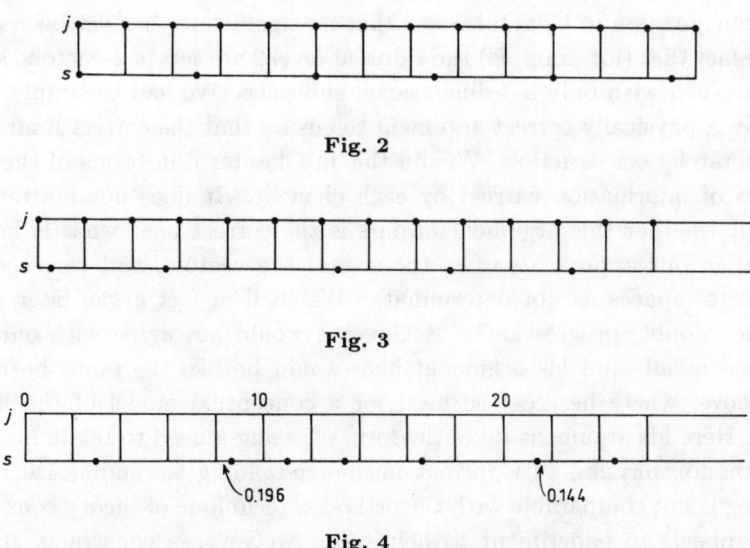

Fig. 2

Fig. 3

Fig. 4

In a situation such as Fig. 4 there is no s_0 because, no matter how small a number is prescribed, it will always be possible to find a part of the two chains for which there is an s nearer to than that to a j. However, it may be sufficient for McGoveran for s_0 to be progressively determined; for example, $s_0 = 0.144$, the value for the eighth value of s and the eleventh of j until the eleventh value is reached, making $s_0 = 0.053$, and this value is then sustained for some considerable number of steps.

McGoveran next relates the idea of two cyclicities to the (for him) particular case of the combinatorial hierarchy. His interpretation of the hierarchy depends critically on certain features which were in the original construction given by Parker-Rhodes but which we have relegated to a less important role in the revised treatment we have given in Chapter 6. This is a second important difference between his treatment and ours, and so we pause to describe it in a little more detail. In the analysis of the hierarchy construction one has, in the original form:

LEVEL	NUMBER OF DCSS OF VECTORS	OPERATORS
1	3 each of 2-vectors	2×2 matrices
2	7 each of 4-vectors	4×4 matrices
3	127 each of 16-vectors	16×16 matrices
$\vdots$		

Now our position in Chapter 6 was that no significance has yet been attached to the fact that (for example) the 7 dcss at level 2 are sets of 4-vectors, since one is concerned with only a 3-dimensional subspace. We had to supply, in place of that, a physically correct argument to ensure that the correct limit arises in the hierarchy construction. We did this in Chapter 6 in terms of the number of bits of information carried by each element. It does not matter for the present whether this argument of ours is the correct one: what is important is that in our approach so far, the size of the vectors, and correspondingly the vector spaces, is not determinate. We shall in fact argue later the need for the "double progression". McGoveran would not agree with our present position at all, and his argument here would be like the point he makes in (d) above, where he sees the need for a conceptual model of the hydrogen atom. Here his argument takes the form of seeing a need to relate his calculus to orthodox physics. To prevent misunderstanding we emphasize that this relating is not comparable with the orthodox technique of theory construction, with appeals to experiment to help out. McGoveran constructs structures in his Ordering Operator Calculus. Then he points out the similarities with structures in physics. He sees the possibility of doing this by saying that the basis representation in terms of vectors provided a *label* for distinguishable objects (though not in exactly the sense in which we have been using the word 'label'). It is then no coincidence for him that the elements inhabit a vector space. Such vector spaces — he would say — are the germs of the vector spaces that abound in physics. He takes this interpretation further by assuming that all of the 4-vectors, not just 7, are there for use. (Strictly not all of them but 15 out of the 16, since he, like us, disregards the zero vector.) Similarly at the higher levels.

McGoveran's assumption (c) then involves one necessarily with a hierarchy up to level 3, and one in which 127 basis strings have been generated, one of which represents j. A subset (attribute) is then at level 2 and so is labelled with one of the 15 strings there. His theory requires a one-to-one mapping of the 127 basis strings at level 3 into the 15 label strings at level 2. One then asks for the probability of coincidence of element and label, that is, for a failure to obtain the required cyclicities, for another cyclicity s, and this is evidently

$$2e = 1/(15.127)$$

(using the notation $2e$ merely for convenience later). McGoveran then returns to the notion of s_0 which we described above and notes that there are two ways in which s and j can fail to start together, viz. by s beginning earlier or starting later than j. This duality reduces the failure probability to e. The basic probability is then, he argues, reduced by confining it to the successful cases, so that it becomes $(1/137)(1 - e)$, and this replaces 137 by 137.035958 which, comparing it with the empirical value of 137.03599 ... is in error by less than 0.00003%, and so is getting on for three times better than Wyler's result.

We turn now to our own calculations and begin by drawing from parts of McGoveran's argument. The first of these is to adjoin the notions of frequency and probability. We differ in where we locate the corrections to the "bare" value 1/137. This value, we say, would occur in nature if we could be assured of a uniform construction through the hierarchy levels. In fact we have to allow for a minimal ambiguity in fixing the level, and this alters the experimental value by an amount which we shall calculate. This minimal ambiguity is part of a much wider kind of freedom which it is essential to exploit in order to get nearer to physics. We shall eventually need to install the possibility of moving about in the combinatorial space, so that processes can be repeated and incorporated into other processes; but this makes no sense unless we have a record separate from the initial process itself. The ergodic principle which has got us off the ground has now to be replaced by something with more structure. When that has been done we shall have taken the first step in defining 'time' and shall have a degree of independence of level changes. Naturally this provision is a far cry from the public absolute time of classical physics, but then no modern physicist would want to go *all* that way.

Let us remind ourselves how the bare value arises. One begins with an artificial constraint, viz. that the system is constrained to operate at the first three levels only. Each entity is then one of 137 possible ones, and the probability, since the three levels are full, of a new entity being any particular one is 1/137. This probability exhibits α arising within this constrained system.

This calculation serves to fix the interpretation of one particular scale-constant. This is important because now the logical position changes. We are no longer in the position of having to justify the scale-constant in terms of $hc/2\pi e^2$; that has been secured because of the substantial agreement with the experimental value. This step uses up, as it were the credit we have built up by finding the number 137, but in return we have lost all ambiguity about what the number is. We can therefore now go on to calculate the corrections to α, knowing that we are dealing with the same number as the physicists.

The second of the strands of McGoveran's argument we call on is his concept of cyclicity which is obviously closely related to the frequency of occurrence of elements in our theory. New elements added to the construction at each stage give information about how the world is. They embody the empirical input, and when we can infer some structure to their appearance we may speak of corresponding *objects*. The *frequency* of elements which constitute such an object is the first source of information about it. To determine such a frequency, the process must label the elements and also keep a record of the number of them. (For the moment, 'frequency' is used in the colloquial sense as when we speak of the relative frequency of encountering a black as compared with that of a white cat on a walk, and not as in the frequency of a pendulum.) In the analysis of the last chapter we paid attention only to the labelling: now we take up the question of how to record in accessible form.

Let us use an ordered pair (x, y) to denote the successive appearances of x, where y is a natural number denoting the order of appearances of x. Thus successive appearances of x will be written $(x, 1), (x, 2), (x, 3), \ldots$. Such an assumption raises questions about our earlier formulation of the basic process. The new entities (ordered pairs) have to be subsumed in some way under the old scheme. However this is in fact done, we shall suppose that it provides some way of giving sufficient accessibility to x. We shall have to discuss later how this accessibility can be provided in the hierarchy by the elements of the algebraic structure. So as not to interrupt the argument we suppose it done. That is to say we suppose that some way is found of placing each x in a space of adequate properties to give sufficient accessibility. This will be called the *process-time* or *p-time*.

To talk about the frequency of an entity is already to jump several guns. Other entities are appearing as well, and it is only against the background of these other entities that the frequency can be defined. Thus the frequency will be defined as a relation between the appearances of different entities. We shall begin with only two such entities for simplicity, but it must be understood that the process will normally be dealing with a large number. If the successive appearances of 1 give a p-time, what about the appearances of 2, 12, etc.? There must be a relation between the appearances of the different entities, and this relation will be the same whether it is exhibiting a p-time or a frequency. Setting up such a relation must be done by the process "considering" sets of appearances of entities. Any meaning we give to 'considering' will involve changes up and down in level, but beyond that we have yet to specify the mechanism.

As we have explained we follow a different line from the third strand in McGoveran's argument in which he uses the matrix framework to anchor his calculation. We have however to explain that we cannot do without some such framework, and the double-progression of the hierarchy construction has always been seen as essential to its understanding. The missing piece in our account is that though we have defined a construction completely we have not said anything about how we label paths "navigated" through the resulting structure so as to have anything comparable with the space of familiar physics. We are therefore in the position of having thrown away the facility which the old hierarchy provided without having replaced it. There is therefore no doubt that at each level a mapping space is required — as McGoveran's method assumes. What is in dispute is only the necessity of its taking the matrix form. For the moment we shall be content with the much weaker assumption of p-time, and to explain how the process constructs p-time we pursue the details of some simple examples.

Evidently p-time and frequency are two related concepts which the process will determine together. This is the first indication we have reached of an explanation of the hitherto puzzling fact that conjugate variables, which are a matter of convenience in classical mechanics, come to play such a critical role in quantum mechanics. But this is to jump ahead. *A priori* we can expect three forms of relation that the process may determine between sequences. The first two are the extremes:

(i) equality of frequency,
(ii) absence of any relation,

Between these two there may be one or more
(iii) partial equivalences.

We begin by studying (i) in more detail; then jump to (iii). The possibility (ii) throws the process back onto the ergodic principle and will be seen to lead to the fine-structure constant.

We shall discuss the relations of types (i), (iii) twice over — once in terms of frequencies, and again in the context of the determination of p-time.

Relations of equality

The simplest relation between frequencies that might be found to hold is that of equality so that, whenever 1 appears it is followed by 2 before another 1 appears, and that 2 is followed by 1 before another 2 appears. The entities might, for instance, appear like this:

$$\underline{1}\,\underline{2}\,12\,\underline{1}\,12\,3\,\underline{2}\,12\,3\,13\,23\,\underline{1}\,\underline{2}\,4\,12\,3\,13\,\underline{1}$$

We can abstract from the whole process just the two entities under consideration. In its abstracted form, the situation could be symbolized as 1, 2, 1, 2, 1, 2, ...

Because of the process nature of the theory, a moment's consideration shows that this is a very special case of equal frequencies, for it might be the case that 2 did not appear for the first time at the beginning of the sequence of 1. In the abstracted notation this more general situation of equal frequencies might be:

$$1\,1\,1\,1\,2\,1\,2\,1\,1\,\ldots$$

To express what has been said about the frequencies in terms of p-time, let n_1, n_2 be the recorded counts of the number of appearances of 1, 2 respectively. In the special case discussed first, we have

when 2 appears, $n_2 = n_1$;   when 1 appears, $n_1 = n_2 + 1$.

In either case, $(n_1 - n_2) \leq 1$, so that $(n_1/n_2 - 1) \leq 1/n_2$, or $n_1/n_2 \approx 1$ for large values of n. The two p-times are the same (as nearly as possible in a discrete theory). The corresponding equations in the more general case are similiarly

$$n_2 = n_1 - 4, \qquad n_1 = n_2 + 5$$

so that $(n_1 - 4)/n_2 \approx 1$, which we could rewrite as $n_2 \approx n_1 - 4$. This shows that the concept of p-time already has the property which in physical time is expressed by saying that t and $t' = t + c$, where c is constant, are equivalent time reckonings. So here one can say that, just as in the special case, n_1, n_2 give the *same* p-times. Other relations of a less restrictive sort would be thrown up by the process.

The sequence 1 1 1 2 1 1 2 1 1 2 ... would finish in the same way with $n_1 \approx 2n_2$. One frequency is double the other. It is natural to compare the two p-times in this case with the well known case for physical times t and t' (equal to ct where c is constant) which are equivalent time reckonings (but for a change of unit). We shall say in the case of p-times that they are equivalent, and this points to a different way of talking about the results according as they are expressed in terms of frequency or p-time. Putting the last two results together, one can see that two p-times are to be taken as equivalent when one of them is approximately a linear function of the other. One says "approximately" because the discrete nature of the theory prevents exact linear equivalence (except in trivial cases) between what are just counts.

Relations of partial equality for frequency

The results above have been derived by looking at very simple examples. When the examples become more complicated (and, as we have said they will in practice be very much more complicated) the concept of p-time has to be developed into something which approaches a little nearer to physical time — at least having a measure which can be a rational number instead of merely an integer count. We shall call the further developed concept r-*time*. This development is important because it contains the first appearance of rational numbers. We continue to exhibit what is going on by simple examples.

(i) Consider in the abstracted form the sequence

1 2 1 1 2 1 2 1 1 2 1 2 1 1 2 1 1 2 ...

and suppose for the purpose of setting up the relation that there is some rule governing the sequence. We ask how the process is going to deal with such a situation? We argue that, however it in fact proceeds, the effect will be the same as if it followed one of two procedures which we shall call *internal* and *external* respectively. Eventually each procedure will get used. We begin with the internal procedure, which is as follows: one seeks to set up what we call an internal relation between the two sequences. This is carried out by a two-stage programme, and it is necessary to enter into the details of this programme because it is here that the entry of rational numbers takes place.

The first stage proceeds as follows: the process has to set up a relation between the appearances of 1 and 2, and for this purpose it must have some common notation. If it looks at the sequence from the point of view of the occurrence of 1 then, as well as the integer counts from 1 to 11, it has to take note of the additional entry 2 between 1 and 2, 3 and 4, 4 and 5, and so on. Whatever notation is used for these "intermediate" positions, the system can be assumed (without loss of generality) to behave by labelling them in an arbitrary way as $1\frac{1}{2}$, $3\frac{1}{2}$, $4\frac{1}{2}$, and so on. Nothing metrical is implied by these signs; $1\frac{1}{2}$ means no more than "between 1 and 2", since "being half-way" would have no meaning at this stage. Then we need a notation for the two 1's between 3 and 4, and by the same argument we may use $3\frac{1}{3}$, $3\frac{2}{3}$, with the same absence of metrical significance. The sequence of 1's and 2's taken together now has, in two alternative ways, complete p-times. We shall continue to use n_1, n_2 for these, although we must remember that they are not just integer counts now, but have conventional "fractional parts".

The two p-times provided for the system are equivalent, and so we now seek to set up a linear relation between the times. This relation will only be approximate because of the conventional values assigned to the intermediate positions. It is not necessary to specify how this happens. We could, looking at it from the outside, do it by the method of least squares. However it is in fact done, the results will not differ very much. When the linear relation has been found (to some approximation which will improve when the sample is longer) the p-times will be adjusted in accordance with it. To put it formally, if n is a natural number, the corresponding r-time (call it r) is equal to n. If n is not a natural number, so that it stands for a conventional assigning, and n' is a natural number, n is replaced by r which is the result of the linear transformation of n'. This process sounds a little clumsy, but it will be clear from an example. In the case above a straightforward application of the least squares technique will give $n' \approx 0.605n + 0.147$. This leads to the *equation* $r' = 0.605r + 0.147$, where, for instance, the assigning $n = 1\frac{1}{2}$ is replaced by r equal the value that will give $r' = 1$, i.e. 1.4 which is $(1 - 0.147) \div 0.605$.

(ii) A little more needs to be said about an example of similar appearance which is however essentially different because of the discrete character of the theory. One can see this most clearly by going back to the previous example and asking how it was put together. It was in fact constructed by supposing that the two frequencies were in the ratio 11 : 7. It was this which made it rule-based. The various positions for a 2 in the sequence corresponded to values of $n \times (11/7)$. One could express it this way: the example is of a recursive function. The rule for it is most easily expressed in terms of the primitive recursive function f defined by

$$f(n) = \mu y_{y<2n}[7(y+1) > 11n] \ .$$

Then the rule is that the kth appearance of 1 in the sequence is followed by 2 if k is a value of $F(n)$: otherwise by 1. The whole specification can equally be set up in terms of 2 rather than 1.

But now consider the rule-determined sequence

$$1\ 2\ 1\ 2\ 1\ 1\ 2\ 1\ 2\ 1\ 1\ 2\ 1\ 2\ 1\ 2\ 1\ 1\ 2\ 1$$
$$2\ 1\ 1\ 2\ 1\ 2\ 1\ 2\ 1\ 1\ 2\ 1\ 2\ 1\ 1\ 2\ 1\ 2\ 1\ 1\ 2\ \ldots$$

Here the sequence does not show any obvious regularity but it is determined by the recursive function defined by f where

$$f(n) = \mu y_{y<2n}[(y+1)^2 > 2n^2] \ .$$

This can be analysed in the same way, but although as the sequence progresses the linear relation does not change much and seems to settle down, it does not finish with any fixed value. This is because the example has been made up with the frequency ratio being $\sqrt{2}$ and that is not something in the theory.

(iii) Lastly, one must not fall into the trap of thinking that relations will always give fixed ratios of frequencies, even approximately. For example the sequence

$$1\ 2\ 1\ 1\ 2\ 1\ 1\ 1\ 2\ 1\ 1\ 1\ 1\ 2\ 1\ 1\ 1\ 1\ 1\ 2\ \ldots$$

could obviously be rule-determined but would not give any frequency ratio.

The next question that it seems reasonable to ask is, what more needs to be developed about r-times before they can be considered an approximation to the time in conventional physics? One can see a sort of answer to this by jumping for a moment to the corresponding situation in conventional theory where the actual ordering of events was taken (in Leibniz' formulation) to define a time reckoning. In that case, also, one might ask whether this is physical time, and the answer is a clear NO, unless it has that special form which makes Newton's first law true. That answer will not do for us, but it does point to the sort of thing that is needed next and that is to see how in the input to the system an external relation can become evident between totally different r-times. We shall consider this question further in Chapter 9, but now return to the calculation of the fine-structure constant because we have developed the notion of frequency and r-time enough for that purpose.

Absence of any relation

We go to the other extreme case from the one where r-times can be determined and consider the situation where no relation exists between the sequences of the various entities. In that case we are thrown back on the ergodic principle. To explain what is happenning, let us consider the artificial constraint where the system can operate on the first three levels only. We deal with 137 elements only, and so must constantly return to elements which have been considered already, giving rise to the frequencies that were described above. We look in some detail at how this occurs.

To do this, we first explain how we present the process. In the general derivation in the last chapter, the relationship between the levels took the form that the element u at the lower level was in the dcs specified by an element A at the higher level if and only if $A(u) = 0$. The elements A consisted of a core and an infinite tail (the latter carrying no information because it was recursively determined by the core). However the infinite tail cannot be ignored completely because it plays a part in determining whether a set of elements is linearly independent. We drew attention to the way in which Parker-Rhodes, in his original derivation, employed a different *belonging* interpretation, using the relation $A(u) = u$. His A was non-singular whereas our core is singular (and can even be zero). The two representations are wholly equivalent as we showed in the last chapter. It will be more convenient to employ the Parker-Rhodes form here because it does away with the need to use the infinite tails.

Now consider the effect of the ergodic hypothesis on the process of arriving at any one of the 137 elements of the hierarchy. The first approximation to this is simply to recognize that a new element may be any of the known ones with equal probability, giving rise to the probability 1/137 and so to use the first approximation of α which identifies the particular constant, as we said above. But at each level the process has not specified whether the element is at that level or not. One needs to apply the ergodic principle again at each individual level. At the first level the possibilities are that the element is 1, 2, 12 or none of these, so with a probability 1/4 of each. In the same way at the next level there is a probability 1/8 of each of the seven elements, or of being at the next level. Similarly at the next level one has a probability of 1/128, and so the probability of not being at any of the three levels but at the fourth (forbidden) is $1/(4.8.128) = e$ (say). The probability of satisfying the artificial constraint is $1 - e$ and so the probability of 1/137, which was produced by applying the constraint, should rather be $(1/137)(1 - e) = 1/(137.033)$. The improvement in the value of $1/\alpha$ is therefore one order of magnitude, since the error is now only 0.002%.

But this argument is inaccurate, as one can see by looking in more detail at the process. Suppose that at some stage the elements 1, 2, 12 have been recalled (not necessarily in that order). The three dcss of the elements 1, 2 can be symbolised at the next level only by the three elements (1, 12), (12, 2), (1, 2). The 7 dcss generated by these will be exactly the same as when they arose earlier in the process. But the 7 elements at the next level which represent each of these dcss need not be the same as in the earlier situation. In fact, as it is easy to see, and will be shown in the note at the end of the chapter, only 61772 do so. 12014 will give 63 dcss, 300 will give 31, and 2 will give 15. The mean number of dcss is then not 127 but 116.23 .

The initial stages of the calculation of the correction to the fine-structure constant are the same as before but the problem at the third level is a more complex one because it is a matter of compound probability. We can shorten the calculation by observing that almost all of the occurrences of the next level (about 8/9) will have 127 dcss, and so it will be a good approximation to take the average number of dcss instead of working out a complex problem. Then one can say that the probability of being at the fourth level is approximately $1/(117.23)$. The probability of not being at any of the three levels and so being at the fourth is therefore $1/(4.8.117.23) = e$ and so the probability 1/137 which was produced by imposing the constraints should be $(1/137)(1 - e)$ which leads to the value for $1/\alpha$ of $137.03652\ldots$, in error by about 0.0004%. Unfortunately this improvement in the result by a further order of magnitude arises through the cancellation of two errors. One error is in taking the mean of the dimensions, 15, 31, 63, 127; what is in fact in play is the reciprocals of them increased by 1. The other error is more serious in principle, though it does not produce any more numerical change. Although it is true and well-known that, given linearly independent elements, the number $n^* = 2^n - 1$, of the elements which they generate is equal to the number of dcss which they generate, this is no longer the case when the elements are not independent. To give a simple example, consider the 1, 2, 3 which generate the usual 7: 1, 2, 12, 3, 13, 23, 123. Now suppose that in fact 3 = 12. Then these seven elements become 1, 2, 12, 12, 2, 1, 0 as expected; the three at the lower level each coming twice + a zero. But the situation with the dcss is different. The original 7 are 1; 2; 1,2,12; 3; 1,3,13; 2,3, 23; ALL. If 3 = 12 however these become 1; 2; 1,2,12; 12; 1,2,12; 1,2,12; ALL in which some of the dcss at the lower level are produced once, one is produced four times, and a new dcs arises. So the numbers of dimensions used in the calculations, which were intended to be the numbers of dcss (later to be represented by single elements at the next higher level) are wrong. It turns out that the correct numbers depend on the number of basis numbers giving rise to a zero. These numbers appear in brackets in the table in the note at the end of this chapter.

It at once becomes clear that the situation with dimensions 4 or 5 is very complicated because of the increase in the number of ways in which the elements can be non-independent. On the other hand the case of dimension 6 can be handled. The value of the approximate calculation is that it shows how unimportant the number of elements of dimensions 4 and 5 are. In fact if they are simply dropped, the value of the mean decreases by 0.13 (or a little less). Now the value of $1/\alpha$ involves this mean, M say, in the form $137/\{32 \times (M+1)\}$ so that a small change δM in M causes a difference in the final answer of $(137\delta M)/\{32 \times (M+1)^2\}$ or about 0.00004 when $\delta M = 0.13$. One can guess that the effect in the corrected calculation will be of the same order of magnitude. In this correction we must preserve the same number of dimension 7 spaces, and then we suppose the rest to be of dimension 6 (and so to be 12316 in number). Now the table of results shows that the original number of zeroes, 12928, was made up of 216 with 3 elements making a zero, 9000 with 4, 12928 with 5, 1600 with 6 and 114 with 7. In our correction we shall assume that the 12316 spaces assumed to be of dimension 6 will divide up in the same ratio, so that (multiplying each figure by the same ratio 12316/12928) they will be 206, 8574, 1903, 1524 and 109. The next step is to calculate the number of dcss in each case. It is easiest to see what is happenning by looking at the general case in which there are N basis elements $1, 2, 3, \ldots, N$ in which N is in fact the sum of r of the basis elements, say $N = 1\ 2\ 3 \ldots r$. One can divide up the set of generators as $1, 2, \ldots, r, (r+1), \ldots, (N-1)$. Then the N^* elements generated consist of:

(a) the $(N-1)^*$ old ones not involving N at all:
(b) the $(N-1)^* + 1$ involving N, some of which are new, some old.

They will be old if they are of the form $D(X; Y; N)$ where D denotes discriminate closure as usual and X has one of the two forms:

(i) $r - 1$ of the elements making up N.
(ii) All of the elements of N.

In case (i) the reason is that $D(X; Y; N) = D(X^+; Y)$, where X^+ denotes the result of replacing the missing element of N in X, and in case (ii) $D(X; Y; N) = D(X; Y)$. Here Y can be any of the elements from $r + 1$ to $N - 1$ and so (i) can occur $r \times 2^{N-r-1}$ times and (ii) can occur 2^{N-r-1} times. In all then

$(r+1)2^{N-r-1}$ of the elements generated are old ones and so $N^* - (r+1)2^{N-r-1}$ are new. For the case of $N = 7$ we have the following table of new elements and total dcss:

$r =$	2	3	4	5	6
New	16	32	44	52	57
Tot	79	95	107	115	120

These are the correct number of dimensions to use in the corrected formula, each augmented by one because of the probability calculation. So we derive:

$$(206/80) + (8574/96) + (1903/108) + (1524/116) + (109/121) = 12316/D ,$$

and also

$$12316/D + 61772/128 = 74088/E ,$$

where the next approximation to $1/\alpha$ is $137 + 137/32E$, giving the approximation to $1/\alpha$ of 137.03503, in error by less than 0.001%.

The source of error is two-fold. Some of it lies in the approximation in the method of calculation (neglecting the dimensions 4, 5 and supposing that the set of zeros divided in the same ratio) but our argument shows that this is not likely to be very great (perhaps one-tenth of the discrepancy). The rest arises in this way: We have up till now disregarded the operation of the actual hierarchy-building process — asking when it is likely that the process will ascend a level on the grounds that one level is more-or-less completed? A little experiment shows that it is not so trivial as one might think to devise a rule, but whatever the form of the rule that the process is following, it must involve judging in terms of the number of "failures" to get a new element in the discrimination.

Now when one element is not independent, it is clear that the number of such "failures" will be increased and this will lead to an underestimation of the total number of dcss. By how much, is something that depends on the exact form of the rule for level change. But it will decrease E and therefore increase $1/\alpha$. So it is clear that the calculation has given a lower bound to the value of $1/\alpha$, and that the exact value will depend on the nature of the process, and therefore on the exact experimental situation under which the fine-structure constant is determined.

Although some work remains to be done on the exact value of the fine-structure constant, we have taken a significant step forward in this chapter. We have shown how an experimental value of one of the scale-constants may be calculated in the theory, although it does not have an integral value. In this way we have made it clear how the numerical frame of the hierarchy can be regarded as part of a wider field of numbers with physical interpretations, and this understanding is of more importance to our argument than the substantial numerical agreement of the fine-structure calculation. It has been possible to reach this point by considering r-times for only two elements and the internal relations between them. The introduction of more than two r-times, and the external relations between them, which we shall take up again in Chapter 9, will lead to the construction of space.

Note to Chapter 7

We have to find the dimension of each of the vector spaces generated by each set of 7 operators corresponding (as we said in the text) according to the Parker-Rhodes convention to the 7 dcss at the lower level. There are 74088 such sets, as is clear from the beginning of the calculation below. The rest of the calculation falls into two parts. In the first part we begin by showing that the dimension cannot fall below 4 and we calculate the number of cases of dimension less than 7 by first counting the number of ways in which a sum of operators can be zero. The total number of zeros is 12928. In part 2 we tackle the rest of the problem; the spaces of dimension 6 have a single zero; those of dimension 5 have 3 and those of dimension 4 have 7 (because the set of zeros is a dcs). It is necessary to begin by counting the number of spaces of dimension 4, 5 (2, 300 respectively) because this is not too difficult. These account for $2.7 + 300.3 = 914$ of the zeros, leaving 12014 which must be of dimension 6.

The number of sets is 74088

Consider the most general set of operators for the seven dcss generated by the three elements at the second level. Take these three, without loss of generality, as 1, 2, 3. The set of operators is

$$
\begin{array}{lll}
(1, a, b) & (g, 2, 3) & I \\
(c, 2, d) \quad \text{singlets,} & (1, h, 3) \quad \text{triplets,} & (1, 2, 3). \\
(e, f, 3) & (1, 2, i) &
\end{array}
$$

In this table, a, b, c, d, e, f, g, i may have any of the values 1, 2, 12, 3, 13, 23, 123 (using here and later the abbreviated notation xy for $x + y$, where $+$ is discrimination). But not all of these values can be taken simultaneously. In fact, the set of operators is a set of generators for a vector space u, and there are 74088 such spaces, forming a set U, say, of vector spaces. This number arises as follows: Consider first the operator $(1, a, b)$ and calculate its effect on the set of elements:

Elements	1	2	12	3	13	23	123
Effect	1	a	$1a$	b	$1b$	ab	$1ab$

The only eigenvector allowed is 1 and none of the other effects may be zero. Accordingly, $a \neq 2, 1$, $b \neq 3, 1$, $ab \neq 0, 1, 23$. This may all be summarized in a table:

$a\backslash b$	2	12	13	23	123
12				13	3
3		123		2	12
13	123			13	2
23	3	13	2		
123	13	3	2		

Here the entries in the table are the values of $a + b = ab$ if these are allowed by the restrictions. It will be observed that there are just 14 possible pairs a, b. By symmetry the same is true of c, d and e, f. The same treatment of $(g, 2, 3)$ will show that $g \neq 1, 2, 3, 23$ and so can have one of the three values 12, 13, 123. In all then there are $14^3 . 3^3 = 74088$ such spaces.

PART 1. The number of zeros is 12928

The purpose of the calculation is to determine how many of these 74088 spaces are of dimension 7; how many of dimension 6 and so on. Consider any space u as a direct sum of two spaces, u_1 being the space generated by singlets alone, and u_2 being the space generated by the other four, and let the corresponding sets be written U_1, U_2. Thus the number of elements of U_1 is $14^3 = 2744$ and that of U_2 is 27. Now the dimension of any u is at least 4 because dim $u_2 = 4$. This is most easily seen by listing the 15 elements of any u_2 and since this list will be needed later the elements are numbered. The elements 1, 2, 3, 4 are linearly independent and generate the whole space; for the restrictions on the values of g, h, i ensure that none of the entries in the table can be zero (for the only possibilities are 5, ..., 10 and 15, and $g \neq 1$, $h \neq 2$, $i \neq 3$ rules these out). The problem is therefore restricted to determining how many

1.	1	2	3
2.	g	2	3
3.	1	h	3
4.	1	2	i
5.	$1g$	0	0
6.	0	$2h$	0
7.	0	0	$3i$
8.	$1g$	$2h$	0
9.	$1g$	0	$3i$
10.	0	$2h$	$3i$
11.	g	h	i
12.	1	h	i
13.	g	2	i
14.	g	h	3
15.	$1g$	$2h$	$3i$

spaces are of dimensions 7, 6, 5 or 4. It is easy to determine those of dimension 4, and not too difficult to do the same for dimension 5, but the large number of ways of being of dimension 6 easily causes errors. To avoid these errors we go over to the method of first counting the zeros, then determining which are seven-fold zeros, corresponding to dimension 4, which are triple, corresponding to dimension 5 (since these are both straightforward to find). The remaining zeros must then be single ones, corresponding to dimension 6.

The seven generators of a space u give rise by discrimination to 127 elements, and if the dimension of u is less than 7 some of these elements will be zero. If x in $U = U_1 \oplus U_2$ is written $x = x_1 + x_2$ with x_r in U_r $(r = 1, 2)$, then by symmetry it will be possible to limit attention to the sample values of x_1 : $[(1, a, b), (ce, 2f, 3d), (1ce, 2af, 3bd)]$ so long as we use all the possible values of x_2 and also of 0. Some of the numbers of counts will then need to be multiplied by 3 to take account of the symmetries. The result of all this is listed in the table below, where the separate cases to be investigated are numbered off, with a decimal added for the column. The numbers in the table are the results of the following calculations, inserted here in anticipation: those in the first two columns being the ones that need to be multiplied by 3. The numbers in the brackets should be ignored at this stage.

Before going into the details of the calculation, it is desirable to tidy up the notations (to avoid as far as possible any errors arising from clumsy notations). This tidying up uses the idea of symmetry mentioned above. Let the three operators P, Q, R be those of interchanging the base vectors of the original space on which the operators act, P being $2 \leftrightarrow 3$, Q, $3 \leftrightarrow 1$, and R, $1 \leftrightarrow 2$. In the table for a, b, both the elements were listed in order of magnitude in the usual ordering.

Table of elements of any u. It is easy to see that no zeros can arise in the squares where no numbers are entered. The groups of three equal numbers arise at once from symmetry. The six occurrences of 198 are two triples which are equal from a slightly more general theorem about symmetry, but it is not worth detailing.

				1	a	b	ce	$2f$	$3d$	$1ce$	$2af$	$3bd$
0.										216	(3)	
1.	1	2	3							216	(4)	
2.	g	2	3							198	(4)	
3.	1	h	3							198	(4)	
4.	1	2	i							198	(4)	
5.	$1g$	0	0							198	(5)	
6.	0	$2h$	0							198	(5)	
7.	0	0	$3i$							198	(5)	
8.	$1g$	$2h$	0							132	(5)	
9.	$1g$	0	$3i$							132	(5)	
10.	0	$2h$	$3i$				378	(4)		132	(5)	
11.	g	h	i				168	(5)		112	(6)	
12.	1	h	i	2352	(4)		168	(5)		174	(6)	
13.	g	2	i							174	(6)	
14.	g	h	3							174	(6)	
15.	$1g$	$2h$	$3i$				322	(6)		114	(7)	

It is more convenient here to list the basis member, 3, of a first and then the others in order. Having done that, the way of listing the others should be determined by the symmetries: $b = Pa$, $c = Ra$, $d = Rb$, $e = Pc = PRa$, $f = Pd = PRb$. In this new ordering, we reconstruct the table for a, b on the first page of this note in the form of the characteristic function f for a, b, where $f(x, y) = 1$ if x, y is one of the 14 possible pairs of a, b and zero otherwise. With this ordering the new table has a symmetric appearance and, what is more important, the same table arises for the other pairs:

	$e \backslash f$		1	23	13	12	123
	$c \backslash d$		1	23	12	13	123
		$a \backslash b$	2	13	12	23	123
2	3	3	0	0	1	1	1
13	12	12	0	0	0	1	1
23	23	13	1	0	0	1	1
12	13	23	1	1	1	0	0
123	123	123	1	1	1	0	0

The table for e, f is that using the top line and the left column; that for c, d uses the middle ones in each case.

It is unnecessary to set out all the calculations. We begin with 0.3 as an easy one, and then 2.3 as a typical one. We use 11.3 to show the need for a more subtle approach which we then apply to the other cases in more or less detail.

0.3. The conditions for a zero are $ce = 1$, $af = 2$, $bd = 3$. We first simplify the calculation by noting that if $ce = 1$ it is impossible that $e = 12$ or $c = 13$. By using the relations between a, b, c, d, e, f in terms of P, Q, R it follows at once that $a \neq 23$, $b \neq 23$, $d \neq 13$, $f \neq 12$ and so the fourth line and the fourth column of the table for the characteristic function can be struck out in this case. Using the reduced tables allows one to set out the possible cases:

a	b	$d = 3b$	c	$e = 1c$	f	$2a$
3	12	123	23	123	23	23
12	123	12	123	23	1	1
13	2	23	123	23	123	123
13	123	12	3	13	123	123
13	123	12	123	23	123	123
123	13	1	23	123	13	13
123	12	1	12	2	13	13
123	12	1	23	123	13	13

The table exhibits the eight possible cases. Other possibilities from the tables run into inconsistencies. In this case, g, h, i are not restricted, so that this result must be multiplied by $3^3 = 27$, giving 216, as stated in the table.

2.3. The conditions in this case are $ceg = 1$, $a = f$, $b = d$. Since $a = f$, the values $a = 3$, $f = 1$ are ruled out and similarly $b = 2$ and $d = 1$ are excluded. A characteristic function for (c, e) can be tabulated, since $1g = 2, 3, 23$.

Scattering and Coupling Constants

	2	13	23	12	123
3	1	0	1	0	0
12	0	1	0	0	1
23	1	0	0	0	0
13	0	0	0	1	1
123	0	1	0	1	0

Using this table and the reduced tables for the other variables, the possibilities can be listed as in Fig. 5, giving rise to an obvious 7 cases when $c = 3$. Each of the other four cases of c will be found to give rise to four solutions, except one in which only three are possible, so that the total is 22. In each of these cases g is already determined but h, i are free, so that the total must be multiplied by $3^2 = 9$, giving 198 as stated.

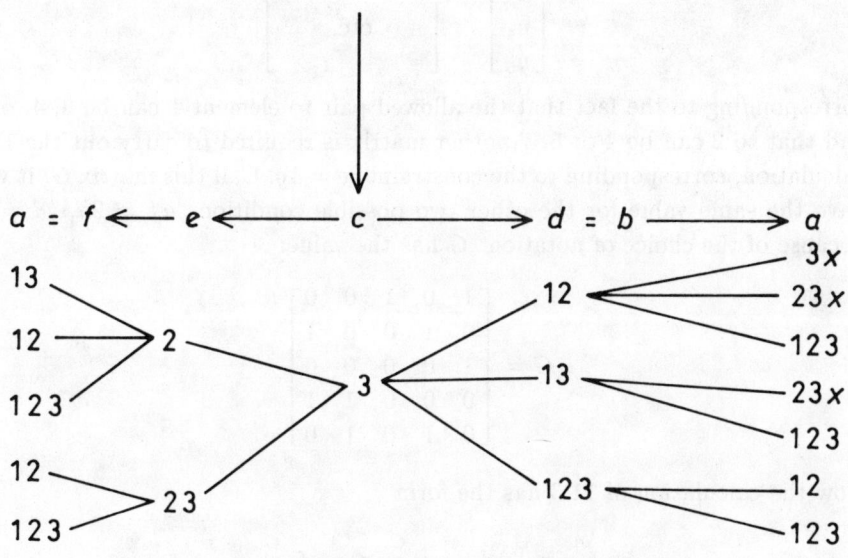

Fig. 5

11.3. Here the conditions are $ce = 1g$, $af = 2h$, $bd = 3i$. The table for c, e used in 2.3 will again be useful here, but the same table will also hold for a, f and b, d so long as these are listed as described. This is because applying Q turns c, e to d, b and also turns the restriction $ce = 1g$ into $db = 3i$. The reader will easily see that the same technique that has just been used for 2.3 will give rise for $a = 3$ to a graph with 21 nodes and 35 joins. The number of paths in this diagram can be found by the standard network theory to be 28 but it is clear that the method is in danger of error. Accordingly we go over to a better one, partly founded on it, which we call the matrix numerical method. The first step in this method is to work with a new numbering of elements; use the numbers $1,\ldots,5$ for the five allowed values for a,b,c,d,e,f, but according to the lists above. Thus 2, for instance, does not always refer to the same element but denotes 12 when it refers to a or c, 13 when it refers to e or b and so on. Only 5 has always the meaning of 123. Then the table for allowed combinations can be viewed as a matrix, say A (for 'allowed'). (However note that it is a matrix whose elements are from Z, NOT Z_2). Its use as a matrix is in this way:

$$A \begin{bmatrix} u_1 \\ u_2 \\ u_3 \\ u_4 \\ u_5 \end{bmatrix} = \begin{bmatrix} u_3 + u_4 + u_5 \\ u_4 + u_5 \\ \text{etc.} \end{bmatrix}$$

corresponding to the fact that the allowed pair to element 1 can be 3, 4, or 5 and that to 2 can be 4 or 5. Another matrix is required to carry out the 11.3 calculation, corresponding to the constraint $ce = 1g$. Call this matrix G; it will have the same value for the other two possible conditions $af = 2h$, $bd = 3i$ because of the choice of notation. G has the value:

$$G = \begin{bmatrix} 1 & 0 & 1 & 0 & 0 \\ 0 & 1 & 0 & 0 & 1 \\ 1 & 0 & 0 & 0 & 0 \\ 0 & 0 & 0 & 1 & 1 \\ 0 & 1 & 0 & 1 & 0 \end{bmatrix}$$

Now the calculation of 11.3 has the form

$$a \xrightarrow{A} b \xrightarrow{G} d \xrightarrow{A} c \xrightarrow{G} e \xrightarrow{A} f \xrightarrow{G} a \,.$$

The network for 11.3 can be described by beginning with a and operating on it successively with A, G, A, G, A, G. Then the result, $(GA)^3 a$, is constrained to be a and the question in which we are interested is in how many ways this is possible. Constraining the result to be a means looking at the diagonal elements of the product matrix, and since we are to look at the effects on all five elements and add up the numbers, we are seeking the trace, $\text{tr}(GA)^3$. This is easily found to be 112, which is the result given in the table (and is more than the 28 given above because 28 was the number of paths through the one graph for $a = 3$ and to this must be added the number for each of the other allowed values of a).

It is instructive to see how this matrix numerical method works for the two results already calculated. Consider first 2.3, for which the equations to be solved are $ce = 1g$, $a = f$, $b = d$. The inequalities necessitate the introduction of a third matrix, E, for equality. (This is the price to be paid for using the numerical notation.) In the old notation, $a = f$ means

	a	$=$	1	2	3	4	5
		$=$	3	12	13	23	123
	f	$=$	*	12	13	23	123
		$=$		4	3	2	5

and so for the other pairs of variables, so that the matrix for E must give nothing for 1, interchange 2 and 4 and leave 3 and 5 unchanged; that is, it is:

$$E = \begin{bmatrix} 0 & 0 & 0 & 0 & 0 \\ 0 & 0 & 0 & 1 & 0 \\ 0 & 0 & 1 & 0 & 0 \\ 0 & 1 & 0 & 0 & 0 \\ 0 & 0 & 0 & 0 & 1 \end{bmatrix}$$

Then the result for 2.3 is easily seen to be given by tr(AEAGAE) and this is $0 + 5 + 6 + 4 + 7 = 22$ as before. In much the same way 0.3 needs the new operators S (for 'sum') corresponding to $e = 1c$, which has the form:

$$S = \begin{bmatrix} 0 & 1 & 0 & 0 & 0 \\ 1 & 0 & 0 & 0 & 0 \\ 0 & 0 & 0 & 0 & 1 \\ 0 & 0 & 0 & 0 & 0 \\ 0 & 0 & 1 & 0 & 0 \end{bmatrix}$$

and then 0.3 comes as $\text{tr}(AS)^3 = 1 + 1 + 3 + 0 + 3 = 8$.

12.1. It is safest to use the new method here, although it is quite easy to do 12.1 by first principles. Since $a = h$ and $b = i$ the only possible values for a and b are 2, 4, 5 and so the construction of 12.1 is:

$$a \xrightarrow{A^*} b \xrightarrow{A^*} a$$

where A^* is the reduced form of S when the rows and columns for 1 and 3 are struck out. Then $\text{tr}(A^*)^2 = 4$. This leaves c, d, e, f, g free so that the total result is $4 \times 14^2 \times 3 = 2352$ as stated.

12.3. The new method saves a great deal of work in tackling 12.3. The constraints are $c = e$, $af = 2h$, $bd = 3i$, so that one method of construction is:

$$a \xrightarrow{G} f \xrightarrow{A} e \xrightarrow{E} c \xrightarrow{A} d \xrightarrow{G} b \xrightarrow{A} a$$

and $\text{tr}(AGAEAG) = 58$. But g is free, and $58 \times 3 = 174$. Without going into details, it will be found that 5.3 needs one further matrix, H, corresponding to $ce = g$ and this is:

$$H = \begin{bmatrix} 0 & 0 & 0 & 1 & 1 \\ 0 & 0 & 1 & 0 & 0 \\ 0 & 1 & 0 & 1 & 0 \\ 1 & 0 & 1 & 0 & 0 \\ 1 & 0 & 0 & 0 & 0 \end{bmatrix}$$

Then the result for 5.3 is $\text{tr}(ASAHAS) = 22$ and $22 \times 9 = 198$. Similarly for 8.3 the result is $\text{tr}(ASAHAH) = 44$ and $44 \times 3 = 132$. For 10.2 the restrictions $h = f$ and $d = i$ mean that d, f are 2, 4, 5 and this has to be taken with the fact that $c, e \neq 1$. It is easiest to tabulate the results of which there are 9 with a, b, g free, so $9 \times 14 = 378$. In the same way, it is easiest to do 11.2 and 15.2 by direct methods.

PART 2. The number of spaces of dimension less than 7 is 12316

In the case where a zero results from a space having 6 dimensions it is a single zero, and so this gives a rough approximation to the number of spaces of dimension 6. But if the space has fewer dimensions, there will be repeated zeros; for dimension 5 there will be three-fold ones and for dimension 4, sevenfold. The overall dimension cannot be less than 4 because the dimension of any u_2 is 4 (as explained above). It is easy to trace the 4-dimensional cases, and not too hard to do the same for the 5-dimensional ones. It is quite difficult to find the 6-dimensional case direstly, but these can now be found from the total number of zeros when the 4- and 5-dimensional cases are known.

The equations for the 4-dimensional case are (expressing the case that the three basis elements of u_1 belong to u_2): $a = h$, $b = i$, $c = g$, $d = i$, $e = g$ and $f = h$. This limits the values for $a, \ldots, f$ all to the range 2, 4, 5 and then the scheme is

$$a \xrightarrow{A^*} b \xrightarrow{E^*} d \xrightarrow{A^*} c \xrightarrow{E^*} e \xrightarrow{A^*} f \xrightarrow{E^*} a$$

where A^* and E^* denote the matrices limited to these values. Then the number required is $\operatorname{tr}(E^*A^*)^3 = 2$ giving rise to $2 \times 7 = 14$ of the zeros.

There are in principle three ways of producing a 5-dimensional space. Evidently not all three of the basis elements of u_1 can belong to u_2 or we should have the 4-dimensional case. It may be that just one is not in u_2; if so, we choose it as $(1, a, b)$ and multiply the results by 3 to take account of this choice. It may be that two of the basis elements are not in u_2 but one is. We take the one which is in u_2 to be $(c, 2, d)$ and multiply the results by 3; at the same time the other two basis elements cannot be giving rise to new dimensions and so they must differ by a member of u_2. Thirdly, it may be that none of the basis elements is in u_2 though in this case they must differ by a member of u_2. We deal with the three cases in turn.

Case 1. If $(1, a, b)$ is not in u_2 it is clear that the condition is $(a, b) \neq (h, i)$. And if the other two are in u_2 the only way in which this is possible is for $(c, 2, d) = (g, 2, i)$, $(e, f, 3) = (g, h, 3)$ so that $c = e = g$ and therefore c, e can only have the values 2, 4, 5. Because of the inequality which has to be checked it is easiest to set out the possibilities:

d	c	e	f
4 or 5	2	4	2
2	4	2	4 or 5
2	5	5	2

giving rise to $2 + 2 + 1 = 5$ cases. Now the inequality is $(a, b) \neq (h, i) = (f, d)$. The five cases have these values of (f, d) : 2, 4; 2, 5; 4, 2; 5, 2; 2, 2. These give rise to prohibited values of (a, b) of 4, 2; 4, 5; 2, 4; 5, 4; 4, 4. However the second, fourth and fifth of these are not allowed in any case by A so it is only in the other cases that one of the 14 values of (a, b) is ruled out. Thus the number of spaces in Case 1 is $5 \times 14 - 2 = 68$, and when this is multiplied by 3 we have 204 spaces.

Case 2. It will be noticed in the above calculation that matters were much simplified because in making $(c, 2, d)$ a member of u_2 there was only one possible candidate, i.e. No.13, to which it could be equated. In cases 2 and 3 this simplification is not present. The condition for case 2 is: $(a, b) \neq (h, i)$, $c = g$, $d = i$, and $(e, f, 3) + (1, a, b)$ belongs to u_2. We have to ask first which members of u_2 can be $(1e, af, 3b)$ and a process of elimination soon gives only the four possibilities 9, 11, 13, 15 which have to be dealt with in turn. A direct application of the number matrix method in case 9 gives a trace of 4 but it is necessary to tabulate these 4 cases to ensure that h, i are chosen so as not to invalidate the inequality. We have already that $b = i$ so that the inequality simply requires $a \neq h$. It transpires that in two of the four cases there is no restriction on h, and in two there is a single restriction, so that the total number in this case is 10. In case 11 the condition $af = h$ ensures that the inequalities are satisfied automatically, and the trace method gives 6. In case 13, $b = 3i$ ensures the inequalities and the trace is 2; and in case 15 the inequalities are again satisfied automatically and the appropriate trace is 10. Thus in case 2 there are 32 possible spaces and this has to be multiplied by 3, giving 96.

Case 3. This is the most tedious because we have to deal with the two cases $(c, 2, d) + (1, a, b)$ in u_2, $(e, f, 3) + (1, a, b)$ in u_2. The first might happen for numbers 8, 11, 14, 15, of u_2, and the second, as before, for 9, 11, 13, 15. So there are 16 possibilities to consider; the number is reduced a little by symmetry. It transpires that only in one case is it possible to satisfy the equations: 8, 9, but in this case also it is impossible to satisfy the inequalities. In all, then, there are exactly 300 spaces of dimension 5.

The total number of zeros listed is 12928. Of these zeros, 14 are seven-fold, corresponding to dimension 4; $3 \times 300 = 900$ are triple ones for dimension 5, leaving 12014 spaces of dimension 6. The total number of spaces of dimension less than 7, is therefore 12316, leaving $74088 - 12316 = 61772$ of dimension 7 as stated in the text.

CHAPTER 8

Quantum Numbers and the Particle

We have claimed to make our primary contact with experiment through high-energy scattering processes, and now develop this position. High energy physics is a curious subject. It comprises a vast amount of experimental information, but in spite of the great prestige it commands, the theoretical grasp which it provides does not seem all that impressive if one is prepared to stand back and take a long dispassionate look. What we can say at the outset is that there are phenomena in scattering processes which make us speak of particles, and that these particles fall into a variety of different sorts. We can be ultra-careful and avoid the word 'particle' by saying that the sequences of counts fall into different sorts whose distinctness becomes apparent when we attach probability amplitudes to the interactions. It is agreed on all sides that the first step to understanding why there are these different sorts must come from the classes of fundamental interactions with their characteristic coupling constants (including the weak coupling) which were discussed in Chapter 7.

The most prominent thing in high energy physics which strikes one, therefore, is the strangely dispersed series of interaction strengths expressed numerically by the coupling constants. The second thing is the extraordinary differences in ranges of these forces. The electromagnetic and the gravitational interactions only fall off as geometry would lead one to expect — as the inverse square — and one could say they were universal. The rest, including the weak interaction, are local in their effects. The first of these features is quite unexplained by current theory, but we claim to give an explanation. The second is explained in ways which we can absorb into our picture and we will discuss below. In 1938, Wick, developing the ideas of Yukawa, pointed out that because of the relativistic limits, the exchange of massive particles must produce short-range forces. (The range being of the order h/Mc where M is the mass of the appropriate particle for the weak and the strong interactions respectively.) Later work on these interactions incorporated these notions. However it has to be said that the success in understanding the second feature has tended to obscure the absence of any understanding of the first.

High energy physics has developed its own new concepts which were not to be expected from general physics. Three are very striking, and they are concepts which can quite easily be presented independently of the continuum background which the quantum theory assumes. They stand in their own right, which is obviously an important fact for us, even though the usual way to look at it would be to say that since the continuum dynamics of quantum theory underlies the whole of the phenomena the new concepts must all be one with it.

1. The evidence from bubble chambers and similar experimental devices is classified in terms of quantum numbers. Quantum numbers appeared in low energy quantum physics in a strictly mechanical context. Here, their appearance is rather different. It is useful to say they are *descriptors*. 'Descriptor' is a term from computational language studies where one wants to give a list of properties which an object to be classified has and which is a subset out of all the properties such objects might have. A table of properties with 0 or 1 against each then specifies the object. The binary properties are the descriptors. The quantum numbers are a little more complex than descriptors since they are not all-or-none but have two or more non-zero values. Thus charge can be +, −, or 0, whereas a descriptor would be just + or 0 (meaning zero or not zero). This does not alter the conclusion that the quantum numbers are more like descriptors than they are like dynamical quantities, despite their origins, because the multiple case can be replaced by a greater number of binary descriptors. Charge, for example, would be replaced by two descriptors. It has been found necessary greatly to increase the number of descriptors from the original charge and spin, and the new ones have no direct macroscopic counterparts (a development which our view of them as descriptors makes us see as only to be expected).

2. The proliferation of particles and correspondingly of quantum numbers since 1930 delivered a body blow to the view of the particles as the building blocks of matter, even though one still sees that language employed. Quantum field theory had to compromise its traditional view of the separability of the

fields and the particles without saying so and it did this by acknowledging that a field would always be associated with a corresponding particle even though that particle might not be directly detectable or the force have macroscopic effects. When one asked what the new particle spent its time doing, the answer was that it was an exchange particle. It participated in the lives of two other particles. One did not go so far as to imagine it orbiting them, but nevertheless there was an associated energy. Heisenberg wanted the electron in this role, but main credit goes to Yukawa who postulated the meson, forms of which (to cut a long story short) were later discovered. Exchange forces and exchange particles have been the main successful formative influence in the high energy physics of recent times, and their use in schemes of classification has shown considerable predictive power.

The two high points of theory in the last twenty years have been the unification of the electromagnetic and the weak interactions by Weinberg, Salam and Glashow, and the standard model of quarks and leptons (Gell-Mann, Feynman and others). The electromagnetic/weak work followed the lines we have described of inferring particles of appropriate mass to account for interaction forces in the circumstances of whatever is known about their lifetimes and with regard to the fitting of their quantum numbers with the other participants in the interactions. The predicted W and Z particles were eventually found.

3. Mass is different from other quantum numbers; unique. It is said to be quantized because particles of a given kind have exactly the same mass. However it is not a quantum number in the descriptor sense. It is more like a classical variable in having an assignable, though discrete, numerical value. No explanation exists in current physics for the primary mass ratio — that of the proton to the electron. The masses of unstable particles appear in a different light because the exchange idea permits one to correlate masses with the range of the forces for which they are responsible. Since the ranges are known (with varying accuracy) one can estimate the mass. The argument involves an application of the uncertainty principle to get a minimum energy corresponding to the lifetime of the particle. The fact that no such arguments can be applied to the proton and the electron (which are completely stable) means that a general theory of mass is not to hand.

To relate the foregoing general remarks to our theory we can think of an investigation where we have set up apparatus to concentrate attention on particular simplified object systems on the basis of the quantum number scheme (which is what happens in a high energy laboratory). We are presenting a mathematical form which is meant to describe physical reality as long as that is not taken to imply a real background into which it is fitted. Ordinary language always implies surroundings which could equally be known about in principle, whereas in fact a whole range of different investigations would be needed to know more about them — altering the experimental situation beyond recognition. Scattering counts are given meaning only by their context. In other words only what has been given a meaning in the experimental and conceptual set-up of our choosing can be used. This idea is a strange one, and goes with our basic view that the quantum systems describe the limits of our descriptive facility, or of our information system. Moreover the particle which we traditionally imagine we get information about with just one observation has to be seen as a sequential structure which is specified by the whole hierarchy, but not by a pre-existing background.

How do we build up our theoretical terms? We begin with a definition originally formulated by Eddington and since adopted by Noyes:

"The particle is the conceptual carrier of a set of quantum numbers."

This means that particles have no independent existence in a combinatorial theory. The entities said to suffer scattering in a scattering process are not individually accessible — only statistically. We do not deny that counter firings, for example, are unique NO–YES events, but you cannot order a re-run of a process. You can only get a repetition of an event up to a given degree of likeness by waiting for it, and you do not know how long you will have to wait.

Very crudely; particles correspond to strings (binary vectors) in the labelling scheme, while the places in the strings correspond to quantum numbers. Information is primarily of counts of processes. The experimental scattering situation includes knowledge of many sorts including geometry and dynamics, but when we look critically at the prevailing classical realism about the particles as continuing spatial entities, everything seems to depend on the definition of everything else. In fact the classical theory-language is no longer usable. We argue that these definitions require a recursion using our whole construction. Decisions certainly have to be made, however about what framework to start with, and to fit the counts into. The quantum-numbers/particle relationship is where we elect to begin.

The view of particles we have taken, according to which quantum numbers, then particles, are where theory construction starts, dynamics and spatial extension arriving later as the construction gets adequate to accommodate them, is evidently in sharp contrast with the current Newtonian kind of realist 'commonsense', in which particles continue to be little balls of stuff in motion. Given our view, we are left having to explain why the products of a scattering process make continuous lines in experimental conditions which can be observed (or reasonably inferred when the lifetimes are too short for them to be seen). These lines, and the conditions in which they arise, provide most of the information from which the quantum numbers of the various products of the scattering are found. Small wonder that physicists think of the participants of scatterings simply as Newtonian particles with a whole list of bits of behaviour characteristic of the high energy scene tacked on, and a list of bits of classical behaviour snipped off. The verisimilitude of this simple way of thinking is overpowering: what is wrong with it?

Let us remind ourselves of the aspects of the quantum process which make the participants different in principle and not merely in practice:

1. finite lifetimes — requiring high energy to be available according to the uncertainty principle;
2. possibility of exchange allowing presence of further particles to be inferred;
3. each process being 'irretrievable': this is really the most important point: information is strictly statistical about types of entity rather than about individuals. We could never repeat an experiment to get further information about a chosen particle.

These aspects strengthen the claim that high-energy physics is best seen as essentially discrete or combinatorial. Having said this however, it remains true that we get records of straight lines in the first place, and deviations from those straight lines in the second place, and that it is from these two sources that we make our classification. If we are not to say that we are observing a particle in motion then we shall have to say that a straight path is a repeated process with no new information, and that a path which deviates from a straight line has at least one more bit of information to provide. If the deviation is more complex then we get more information. This way of looking at the matter is consistent with the way we shall build dimensional structures in Chapter 9. The other piece that will have to be fitted from theory rather than from commonsense comes because as we would usually say a track requires many low energy interactions or subsidiary scatterings to compose it by ionization. We merely note here that something is required. We have no specification of energy as yet.

One wishes to relate strings in the hierarchy with quantum numbers and hence with particles. The strings which are available for this treatment are the labels — that is to say they are in the mapping space, and we have to remember that in the construction process whether a string is classified as a constructive element or as a label depends on the level from which one considers it. (Note the difference between our use of 'label' and that of Noyes in *Program Universe*, Chapter 5.) There is no way in which we can make the label describe an association of quantum numbers as well as describing the behaviour of the association of these quantum numbers, that is to say particle, in, say, a scattering process involving other particles. To see this consider a process involving two particles, 24 and 134, or 0101 and 1011, where the four places in each string specify the existence or non-existence of a certain quantum number, and hence the nature of the particle. To conform to the whole spirit of our approach the vertex scattering must be represented by discrimination, since that is the only operation that has arisen. But if the scattering is represented by discrimination then the third particle in the process will be 123, which is written 1110. Its quantum number specification will be changed, and we can look to experiment to see if this process is plausible. As a matter of fact it seems to be almost impossible to carry through such an identification in a way consistent with the known experimental facts. This is not surprising, for all we have done is to describe a relationship between quantum numbers. We have not, and never could, use the algebra to describe what the particles have done. What we have set up is the beginning of *process* itself, and it would be a muddle to talk as though there must be a particle dynamics already in process. *The appearance of the quantized particles shows that we have reached the limits of the classical world: it makes no sense to discuss their peculiar properties by yet more of the classical dynamics.* The potential muddle becomes actual if we suppose we could represent a dynamical scheme with momenta, velocities and so on using the discrimination algebra and nothing else. It is our opinion that this logical confusion becomes almost inescapable in the current ways of describing high energy physics and is responsible for the impenetrability of the subject to normal study. The difficulty is quite different from the difficulty one experiences in learning other difficult branches of physics.

If we can't embed the particles in a dynamical scheme of the familiar sort, what can we do? The hierarchy algebra provides a method which is deeply integral to the whole approach, and that is through level change. The construction process requires that we map associations (which we call dcss) of elements at one level (say strings of length 2) onto label strings of length 4. We thus get 4 descriptors or quantum numbers. We can either say that we have a more complex particle, or that we have by taking the two levels in relation to each other found a way of representing the thing which the quantum numbers describe separately from those quantum numbers. This is the exemplar for dealing with particles and quantum numbers which we have to work with. The mathematics of this exemplar is little understood as yet, but it is what is needed in developing a combinatorics for the particle patterns.

We shall now illustrate the mathematics required to deal with particles and quantum numbers at level 2. Exactly similar considerations will apply at each level. Hitherto we have thought of dcss directly in terms of their definition, but the clue to connecting them with sets of quantum numbers is provided by this easy theorem:

Theorem. Denote the elements of bit-strings representing hierarchy elements by $b_1, b_2, \ldots$ (i.e. so that 1 is represented by $b_1 = 1$, $b_j = 0$ otherwise; 2 by $b_2 = 1$, $b_j = 0$ otherwise, and so on). Then every dcs is the set of solutions of a number of simultaneous homogeneous linear equations in the b_j and vice versa. (The coefficients in the equations are, of course, from Z_2.)

The truth of this theorem will become apparent in the following discussion. Take as an initial example the dcs $\{1, 2, 1+2\}$ and limit the string length arbitrarily to 3. It is clear that this is the solution set of the equation $b_3 = 0$. Again, for the set $\{2+3, 3+1, 1+2\}$ a little consideration shows that this is the solution set of $b_1 + b_2 + b_3 = 0$. In the same way the one-element dcs $\{1\}$ is the solution set of $b_1 = 0$ and $b_3 = 0$, and $\{2+3\}$ is of the equations: $b_1 = 0$ and $b_2 + b_3 = 0$. In general, when the strings are of length n, a single equation gives a dcs with $(n-1)^* = 2^{n-1} - 1$ members, and a set of k simultaneous equations corresponds to a dcs of $(n-k)^*$ members.

This theorem is the first step in providing a particle/quantum number connection. For example, if $b_3 = 0$, the dcs $\{1, 2, 1+2\}$ is defined and this is represented at the next level by an array. In our formulation, from Chapter 6, such an array might be $(0, 0, 1, 2, 3, \ldots)$. In Parker-Rhodes' simplification, cutting off the infinite tail, this becomes the matrix $(1, 2, 1+3)$. So much is also to be found, implicitly, in Noyes' work.

The next step is less clear and needs more work. If b_3 represents a quantum number, we shall also need to consider the case $b_3 = 1$. The solution set of a set of non-homogeneous linear equations is evidently not a dcs. Noyes (whose approach will be presented in more detail below) evidently came across this problem first in a somewhat different form and he seeks to solve it by what he calls "Amson invariance". We paraphrase his treatment to integrate it better with what has just been said. He considers, for every bit-string $b = (b_1, b_2, b_3, \ldots, b_n)$, its complement $\underline{b} = (\underline{b_1}, \underline{b_2}, \ldots, \underline{b_n})$ where $x + \underline{x} = 1$. Then one can define an associative, commutative co-discrimination operation $*$ by the rule

$$a + b = c \leftrightarrow \underline{a} * \underline{b} = \underline{c}.$$

Obviously the table for $*$ is

$*$	0	1
0	1	0
1	0	1

Then a set S is a co-dcs if, for any two different elements x, y of S, $x*y$ is in S. Thus one co-dcs is $\{2+3, 1+3, 3\}$ and this is just the set of 3 complements of the set $\{1, 2, 1+2\}$ which we considered above. This co-dcs is not the solution set of $b_3 = 1$ however, for that would also include $1 + 2 + 3$. Noyes avoids this difficulty by ruling out from his algebra what he calls the anti-null string $(1, 1, 1, 1)$, by analogy with the way that the null string is ruled out.

Since every co-dcs is the set of complements of members of a dcs, the sum operator at the next level serves to represent both $b_3 = 0$ (as in a dcs) and $b_3 = 1$ (as in a co-dcs), and this is as it should be in one usage of the term 'quantum number' (though we shall see below that they are differently used, and this presents a difficulty for Noyes' analysis). However isospin fits in well if the two values of b_3 correspond to proton/neutron (though in an unfamiliar notation), and if the common matrix representing them both at a higher level shows them as two states of a common particle. Spin does not fit easily. If the values of b_3 correspond to integral/half-integral spin, it would be quite inappropriate to represent both bosons and fermions by one matrix.

There is a treatment which is less artificial than Noyes' because it does not rule out the null vector. We simply observe that a matrix A at the higher level distinguishes a dcs at the lower level, namely the set of v for which $Av = v$. In doing this it also marks out the complement (in the set-theoretical sense and not in Noyes') — that is to say the set of elements v for which $Av \neq v$. If the first set is, for example, $b_3 = 0$, then the second set has $b_3 = 1$. Thus one simply recognizes that a Parker-Rhodes operator A divides the level below it into S, $\underline{S}$, where S is the set of v for which $Av = v$ and $S \cup \underline{S}$ is the whole level. The convenience of this approach over that of Noyes is seen when one considers the use of several bits to make up more complex quantum numbers. Suppose one needs to consider the four cases

$$(b_2, b_3) = (0, 0), (0, 1), (1, 0), (1, 1) \ .$$

If S is the dcs defined by $b_3 = 0$, and T by $b_1 = 0$, then the four cases are, respectively, the elements of:

$$S \cap T \qquad \underline{S} \cap T \qquad S \cap \underline{T} \qquad \underline{S} \cap \underline{T}$$

Only the first of these is a dcs however, and so this approach runs into the same difficulties as Noyes' in cases like integral/half-integral spin.

Noyes,[1] has thought it overridingly urgent to make a bridge to the language and conceptual framework employed by high energy physicists at present, and has therefore used a modified form of the hierarchy algebra in which growing content strings are added onto the strings constructed as in Chapters 5 and 6, to provide for the detailed metrical (metrical and temporal) information. The strings themselves then carry the structure of the quantum numbers, and are said to label the content strings. (This use of 'label' is different from that of Chapter 6.) The change makes it possible for him to revert to the conventional particles obeying individual mechanical rules in an independent spatial framework. In particular he can interpret discrimination operations on strings directly as scattering processes in the literal sense that the transformations $A+B \to C$ represent spatial movements of particles. Some description of this treatment of the hierarchy (program universe) is given in Chapter 5. In the view of the writers, however, it is the breakdown of the classical, 'receptacle' view of space (see Chapter 2) in which one can assemble bits of description from different sources and assign them independently to a particle, which characterizes quantum physics and is an inevitable consequence of our explanation of the discrete aspects of the world. With regard to labels, they are a necessary and integral part of our construction of the hierarchy, and it would be a mistake to tack them on ad hoc. Hence, however nice it would be to engage naturally in high energy physics discussion without any provisos about the meaning of the language, we cannot afford that price. However we think it possible to present some essentials of Noyes' picture, provided we are prepared to paint in broad strokes.

To all the questions about the origin of discreteness and particulate structure, Noyes answers that we must fully appreciate and exploit the consequences of the necessarily discrete nature of observation. He wants to get rid of scale invariance by considering the limits imposed by technology rather than by Planck's constant. His is a different point of view from ours. It results in a different formulation of the theory. From our point of view it is not clear why we should be restricted to making observations that way. If the reply is that the only kinds of apparatus provided by nature are discrete, then that seems to beg the question. Noyes considers his approach peculiarly sensitive to the demands of operationalism, but in fact the operationalist requirement is that we should express our theories in terms which reflect the experimental techniques we use. Whether we are restricted to those techniques, and if so why, are separate questions, and they are the questions we need answers to if we are in this way to explain the existence of absolute units.

Noyes claims that the hierarchy scheme has places for all the quantum numbers which are observed and only places for those (none left over). Having identified the 3 level-1 entities he proceeds to the 7 at level 2, and these are: the left electron, the right electron, the left and the right positron, the left/left gamma, the right/right gamma, and the coulomb interaction. This last we do not call the photon which we regard as fictitious: it is (relativistic) interaction at a distance and does not travel.

Quarks are more complex. This statement conflicts with the popular perception that they are the building bricks of all matter. They are only that in the sense that they require high energy for their confinement which is classically equivalent to their being small, though it is in fact only a classical hangover to expect simplicity and smallness to go together. In Noyes' scheme there are 7 independent quark states, and these arise from the basis for level 3. There are 2 flavours — up and down. One has charge 2/3 and the other charge 1/3, and the baryon number is 1/3. One gets neutron and proton by adding, since charge comes in both signs. Within each flavour one gets 4 states of the quark, and there are 2 left and 2 right as with electrons. This repeats the structure at level 1/2, and the weak/electromagnetic coupling is secured by the same bosons as appeared earlier.

Noyes regards the most important characteristic of a quantum number to be its conservation in processes, and the quantum numbers which are effective at long distances are limited to three: lepton number, charge and baryon number.

The electromagnetic/weak theory started new discussion on the question of why there should be large numerical factors separating the strengths of the interactions. This is obviously an outstandingly important and natural question, but people first show a fixity of approach in supposing that interactions should be of equal strength unless there is reason to think otherwise, and then proceeding to look for that reason. In the thinking introduced by Higgs, the natural expectation was to find symmetry everywhere, and in particular between the interaction strengths, and the differences were spoken of as "broken symmetries". Since symmetry is obviously a 'good thing', and breaking it is a 'bad thing', one should seek for a reason why symmetry is only apparently, and not really, broken. Higgs postulated the existence of a hitherto unsuspected heavy exchange particle (boson) or perhaps a family of them to explain the breaking of symmetry. Such "Higgs bosons", as they are called, would have to have a mass of perhaps 300 proton masses to have the large effect required of them. At the time of writing none has been found experimentally, though search is intense.

Ours is a quite different way of looking at the situation. The strongest part of our theory is the rigid reasoning we are able to give to why the relative strengths of the fundamental fields should have the values they do. Even the weak/electromagnetic unification theory does not do that: it correlates the interaction strengths with masses of particles but leaves the actual values quite mysterious. From our point of view it would be odd to see the different values as indicative of a shortcoming. Even more, we see no reason for more unification since the existence of these different values arise out of a theoretical scheme which is already unified in the strongest sense it could be: namely in being a deductive consequence of the principles of the theory.

Suppose direct evidence were found for the existence of Higgs' boson. Would that invalidate our claim? On no account: such a discovery would reinforce the existing principle that particles correspond to fields using an exchange mechanism, and in our approach, as we shall presently emphasize, the particle and field concepts cannot have the independence that familiar thinking assumes for them.

The main activity in high energy physics is to fit the particles, on the evidence of their quantum numbers and other properties (notably mass), into a scheme or pattern. The most successful scheme is called 'the standard model' and is based on SU3 (special unitary group in three dimensions). It would normally be claimed that it is part of the explanatory success of the standard model that macroscopic or universal space-time is imported via SU3, which appears in the combination U1 × SU2 × SU3. In fact one gets very little out of this spatial connection apart from the three-fold character. Indeed the symmetry which the connection most strongly leads us to expect — namely parity conservation, or the equal applicability of physical laws when right-handedness is changed to left — is found to be broken. From the point of view we advocate, all this is to be expected. The scattering processes which we speak of in terms of particles are the most extreme extrapolation of our experimental knowledge that we can have and are the beginnings of any attempt to map the universe — not things which we insert into an existing dynamical scheme.

The three-fold character of the standard model is totally integral to our hierarchical model, as we have fully explained, though its connection with the three-dimensionality of macroscopic experience is very subtle. So as far as that aspect is concerned, it, and the standard model are on the same footing. The threefold basis is an empirical fact for the standard model and comes deductively in the hierarchy. For the rest, it is a matter of comparing the successes of two different combinatorial schemes, and reference should be made to Noyes who has written comprehensively about the matter, (Loc.cit.). If we can really compete on equal terms with the standard model as far as a deductive account of the high energy structure is concerned then we are in a strong position since we already have the interaction strengths in our pockets. To quote Belloc *The Modern Traveller*:

> "Whatever happens, we have got
> The Maxim Gun, and they have not."

We finally broach another area of work which is of great importance and needs to be taken much further than we are able at present. Of all the topics which arise when we undertake the combinatorial revision this is the least obvious from the conventional point of view. We have to *understand* the origin of quantum numbers (the classical ones like charge and spin in particular). We have been looking at quantum numbers as 'descriptors', but of course they have another aspect in current theory, which is to represent a quantity of which they are the unit, and this seems not to need mention since it is a commonsense consequence of the background dynamics which gives them their classical meaning. We, however, have no background dynamics, and so not merely have we to classify the quantum numbers — we have to discover why there are these quantum numbers in the first place, and to *explain* the peculiar inherence of classical dynamics in each. We shall confine our present discussion to the origin of the spin quantum number. Two-vectors are required to label the dimensions, but because of the 'similarity of position' the labelling is not operable. This ceases to be the case in the further construction of 4-vectors, which define transformations which can properly be said to confer individuality on the dimensions. One could say that the individuality of the dimensions is degenerate until the recursive relation is created, and that then it is possible to speak of the operators at the second level as rotations of axes which are the elements at the first. In fact we have the germ of the spinor calculus here. We can take any two 2-vectors and regard their transformation into one another as a rotation, and we can take them to be 01, 10 without loss of generality since the same results follow with any other pair. In this way one can regard the imaginary terms in the normal form for the spinors as equivalent to the recursive relation.

In a contribution to *Quantum Theory and Beyond*, one of us (Bastin) wrote about the combinatorial representation of spin, before the conceptual apparatus which has been used here was available. The argument then used was that if we were to impose upon the 2-vectors an ordering represented by upwards and downwards in the case of column vectors, then a mathematical property identifiable with half-integral spin would emerge from the asymmetry. Thus we might investigate transformations of the 2-vectors which — for example — had no zeroes in the bottom place. We require two discriminations to perform one of these, whereas one 2×2 matrix will do the trick. The argument then ran that the frequency of the first process would be half that of the second and therefore that any measure of the physical quantity attached to the quantum number (angular momentum) thus halved, giving the familiar half-integral spin. The statistical argument is of a type now familiar to us, and in general one can say that the right ideas were being forced into a mathematics which they only partly fitted — notably at the point where significance was attached to upness and downness of vectors in the absence of a distinction between labels and dcss.

What about all the other quantum numbers? The most pressing is charge, and this can be constructed in precise analogy with spin, using the device of isospin. One hopes that the more complicated quantum numbers which nowadays exist in such profusion have their roots in more complex interrelations between hierarchy levels but these are mostly unknown and more work is badly needed. One may ask whether all the quantum numbers are on the same footing, seeing that the more complex and modern ones do not seem to have classical analogues. We suggest that the difference is only apparent and arises because we have developed a macroscopic language corresponding to levels one and two (classical mechanics and electromagnetism) but no number like charge which has a macroscopic counterpart.

Conventional theory hardly recognizes the need to explain the origins of a quantum number like charge which has a macroscopic counterpart. If there is space — it would be argued if the question were raised at all — there will be rotations and there is angular momentum and it is this which is quantized. However for us the question why, and in what circumstances, classical dynamical quantities correspond with quantum numbers is profound. Moreover the conventional position that there is no problem because the association is entirely natural, is correct in one way. No dividing line can be drawn between the macroscopic description and the level of the quantum numbers. It has been shown that a beam of protons prepared to have a preponderance of one spin, will transfer angular momentum to a lump of lead, though the experimental conditions have to be very refined.

CHAPTER 9

Towards the Continuum

We now describe the problems involved in going towards 'continuum physics' and carry the solution of them far enough to indicate the general approach. When we speak of the continuum, or of continuum physics, we are not stressing continuum as opposed to discrete, but rather that intuitive haptic knowledge derived from our bodily experiences which, it seems, we call on in applying continuum mathematics in conventional physics. We have hitherto included this in 'the classical theory-language'. One's first thought about the passage to the continuum is likely to be of some kind of averaging over a large number of discrete entities, but the 'averaging' image can be misleading. The intuitively obvious senses of 'averaging' entail spatial distribution.

We first contrast the 'platonic receptacle' view of space with the process view which we advocate. In this respect we find that quantum mechanics and classical mechanics have to be treated quite differently, and the latter occupies us in this chapter. We shall hope, however, to show that in the process view the difference becomes a matter of emphasis rather than a clash between incompatible theories, as it is in current physics. The critical step, which was mentioned peripherally in Chapter 7, is the introduction of external relations between r-times. It is this that enables the sequential process to give rise to structure with a three-fold character; so corresponding to the way in which our bodily experiences are always seen as in a three-dimensional space. We find that external relations enable us to make a significant step towards orthodox physics. To link up with current presentations of special relativity we shall find it easiest to compare our development with Milne's 1938 elaboration of the technique later to be known as the k-calculus. Two aspects of Milne's work are particularly important for us: (i) His perception that the usual discussions of special relativity put more weight on the concept of light-signalling than it can reasonably bear: we go on to discuss the post-Milne form of the k-calculus with that in mind. (ii) His treatment of dimension (which is the major defect of his work). We are able to bring in dimensionality in a satisfactory way. We have to amend and develop the notion of symmetry which we used in early work in the fifties as the fundamental characteristic of dimensionality in a combinatorial view of space. Our method sets its own difficulties which we summarize and answer. We close the chapter with some suggestions for further research in this field.

In the introductory chapter we described the 'platonic receptacle' view of space: space is what holds whatever we care to put into it. Of course the receptacle view is deep in our thinking: it seems to be a way of operating physical theory by projecting our most immediate sensuous knowledge of the world onto each theoretical statement — we need to picture what bodily actions we should take to correspond to it before we can understand it. It is notorious that rigorous analytic solutions to dynamical problems — even in classical physics — never take one far, and one has to get further by approximations and extrapolations. (There is no general solution to the three-body problem.) Our attempt to formalize the classical theory-language was essentially an attempt to replace this sensuous correspondence by a formal structure. We have to remember that when we speak of replacing the continuum we are replacing this too. One has to guard oneself against the easy assumption that analytic geometry and the equations of motion will in principle solve all classical problems and that there is no need to fuss. The point to remember is that what is missing is the analytic technique to link up all the detailed exact bits of solution, for this is where we appeal to the theory-language to carry us over. Such a linking mechanism is already available in the discrete picture provided by the hierarchy in the recursive way in which entities at one level group sets of entities at another. Starting this way, there is no need to require that we build continuum physics; that would probably be impossible. All we can ever do is to provide what we find to be needed for each problem as it comes up: a piecemeal approach appropriate to our process philosophy.

Our whole theory is about the origins of the discrete, and we start from discreteness in the quantum context. We are therefore bound to show how the theory describes classical mechanics whose language is not discrete. Our case is that all the results of current quantum theory are combinatorial in origin and we expect to obtain them by continuation of our combinatoric method. Moreover — we claim — all the mathematical principles which are normally thought to be characteristic of the quantum theory (exclusion, the whole theory of wave-functions and differential operators and eigenstates, Poisson brackets and non-commutativity, Hilbert space) take their familiar form only because of the need to reconcile the discrete character of quantum events with the classical theory-language. It is necessary for us to make and sustain this claim if we are to dispense with the mathematical continuum as a pre-condition for physics. Eventually we shall need to show how and why they perform this function, in order to be able to take over from conventional theory such very accurate results as the Lamb shift, though at present we are at too early a stage in our understanding of spatial relations to do this.

In the formulation of the concept of r-time in Chapter 7, we used for the sake of simplicity 2-cyclicities only. We now want to say a little more about this, preparatory to explaining how the method can be extended to other cyclicities. Firstly we remark that although we used the two entities (1, 2) alone, the process will throw up the discrimination between them, 12 when they come up. Thus the r-time constructed is not for 1, and 2 alone, but for the dcs $D(1,2) = [1,2,12] = S$ say. We can denote it by r_S. Any r-time is constructed from a dcs and any dcs is potentially able to yield an r-time if the necessary construction arises.

Now consider the problem of extending the concept of r-time to include further cyclicities. There are two possibilities. The first may be exemplified by considering say the occurrence of an element 3. The construction is much the same as before. One can begin (and something equivalent will arise eventually anyway) by taking the r-time r_S for elements of S and give values for the p-time of 3, n_3, by a conventional rule as before:

(a) If there are k_3 3's between r'_S and r''_S, give them the conventional values of r_S-time:

$$r'_S + \{u/(k_3 + 1)\}(r''_S - r'_S), \qquad (u = 1, 2, \ldots, k_3) \ .$$

(b) If k_S points of the sequence defined by S lie between n'_3 and $n''_3 = n'_3 + 1$, give them the p_3-time values

$$n'_3 + u/(k_S + 1) \ .$$

This produces the whole sequence labelled, in two ways. Since the two should be equivalent, seek a linear relation between them in any way. (In analysing the system from outside, one might use least squares, as before.) Call this linear relation $r \approx L(n_3)$. Then redefine an r-time by the rule:

$$r_T = L(n_3) \text{ if } n_3 \text{ is integral}, \quad r_T = r_S \text{ otherwise.}$$

The construction has been for the set $S \cup [3]$, but from what has just been said we get r-times for dcss, so that we have derived an r-time for $S' = D(S \cup [3])$. In the same way, if two r-times r_S, r_T have been constructed separately, the elements of T may be taken over one at a time until a new r-time for $D(S \cup T)$ has been constructed. This works so long as S, T are at the same level. This "internal" procedure fails if S and T are at different levels, and the corresponding relation between r_S and r_T is then what we shall call an "external" one.

If the internal procedure is possible and if one makes the

Definition. The change $r \to r^* = f(r)$ where f is monotonically increasing, is called a regraduation of r-time.

Then one has proved the

Theorem. If r_S, r_T are two r-times, r_T may be regraduated so as to be equivalent to r_S.

Astonishingly, this result was anticipated from a different point of view by E. A. Milne.[1] Although Milne's philosophical basis is more profound than the conventional approaches to special relativity, it is still very far from ours. He considers what he calls two "time reckonings" A, B which he defines in terms of readings of "clocks" — a term more general in Milne's usage than in the conventional. He defines a relation called *congruence* between t_A and t_B by means of "observers at" each clock, and "signals" between them. What Milne calls "the first problem of time-keeping" is solved by his theorem that if A, B are not congruent then they can be made so by a clock regraduation (as just defined). Because he starts from a different position from ours, the proof[2] of the theorem is a mathematical one. Our beginning is from a much more detailed analysis of r-times, so that the result follows more directly. From our point of view, the burden of Milne's proof is that it confirms that his initial notions of time-reckonings, clocks, observers and signals are coherent.

Our explanation is not given in terms of these four concepts. We have avoided the anthropomorphism of "observers", and we have constructed r-times without explicit mention of clocks. There remains amongst Milne's

Towards the Continuum

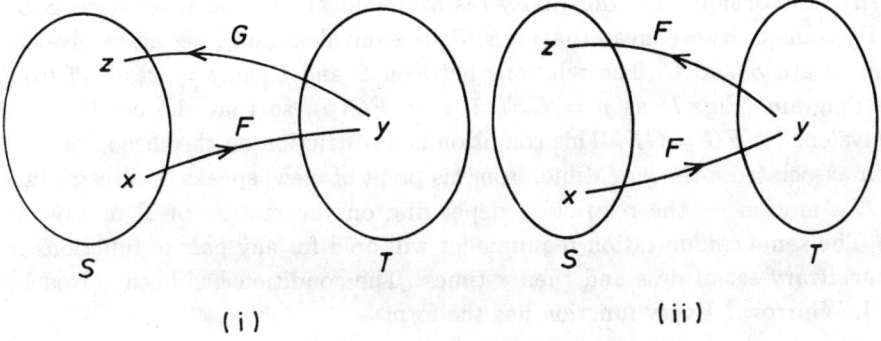

Fig. 6

primitives the idea of light-signals or photons travelling. These have no part in our theory. The discomfort of them is already apparent in conventional special relativity because of the zero proper time which elapses along a "photon path". We see light signals as simply a way of assimilating relations between r-times to the classical theory-language.

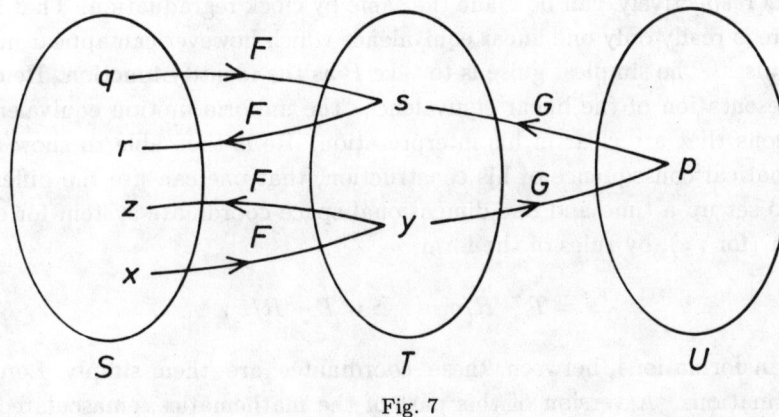

Fig. 7

We return to consider the impossibility of external relations that can arise in the procedure. Here again Milne gives a useful clue for he is led on to what he calls "the second problem of time-keeping", which is the question of whether his definition of congruence (our "equivalence") is, as its name implies, a transitive relation. To act on Milne's clue in our context, it is necessary to set up a little formal theory. The procedure described above sets up a 1:1 mapping between r_S and r_T as shown in Fig. 6(i). If r_T has been regraduated so as to be equivalent, then the relation becomes symmetric, and so $F = G$ as in (ii). To consider the transitivity (as Milne does) we need three dcss, S, T, U. By transitivity we mean that, if S, T are equivalent and T, U are equivalent then so are S and U. The relations between S and U may be read off from the diagram (Fig. 7) as $p = GF(x)$, $r = FG(p)$, so that the condition of equivalence is $FG = GF$. This condition is a restriction on the three dcss and their associated r-times. (Milne, from his point of view, speaks of a restriction on U's motion — the restriction depending on the motion of T relative to S.) The same commutation requirement will hold for any pair of functions in an arbitrary set of dcss and their r-times. The condition has been solved by G. J. Whitrow.[4] Every function has the form

$$F(x) = P[k_S P^{-1}(x)],$$

where P is a monotonically increasing function and k_S is a constant corresponding to each dcs. The details are all given in Milne (loc.cit.). Milne calls such a set of "observers" (in his language) a linear equivalence. The function P determines which *linear equivalence* is under consideration. He is then able to prove his "main theorem" that two linear equivalences, defined by functions P and Q respectively, can be made the same by clock regraduation. That is to say there is really only one linear equivalence which however can appear under many guises. The simplest guise is to take P as the identity function. He calls that presentation of the linear equivalence "the uniform motion equivalence" for reasons that are clear in his interpretation. He is then able to show as a mathematical consequence of his construction, that one can use the different times to set up a time and one-dimensional space coordinate system for each observer (for r_S), by rules of the form

$$x = T - R/c, \qquad z = T + R/c.$$

The transformations between these coordinates are then simply Lorentz transformations. A version of this part of the mathematics, emasculated by removal of much of Milne's important philosophical basis, has now become well-known under Bondi's title — "the k-calculus". The quadratic form, which ought to cause surprise in presentations of special relativity then arises in a natural way because the invariant $dT^2 - dR^2/c^2$ is identical with the invariant $dx.dz$. We can simply take over the main mathematical details of this but its surprising physical significance will be discussed below. In a somewhat obscure passage, Milne goes on to argue that "No quantitative properties of light have been assumed whatsoever." Later he reiterates: "I repeat that only the most primitive elements of perception have gone into the establishment of the Lorentz formulae; and they are in fact the expression of the analysis of the act of perception into its elements. We have not found it necessary to assume the constancy of the speed of light, though this is an *a posteriori* consequence of our analysis."

There are several things that it is important to say about this position of Milne, because we should like to see it as complementing our work. Firstly it is clear, and has often been remarked upon since the publication of his book, that his claim not to be assuming the constancy of the speed of light is as hard to justify as it is to refute. His whole argument is so shot through with appeals to "observers" who "see" other observers' clocks that the exact assumptions are unclear. Why, one asks, are these two terms not just as much in need of fundamental logical reformulation as is the velocity of light? By assuming that the intuitive ideas of them are good enough he argues in a circle. Since we *have* undertaken that reformulation we may claim to be in agreement with him so far as the difference in points of view permits, since no notion of light or of signalling has yet entered our discussion.

Secondly there is the problem of preferred inertial frames in special relativity. It has always been a puzzle that the normal treatments seem to have to start, to quote Einstein, with "a frame of reference in which Newton's laws hold in their usual form". He might have added in, Maxwell's equations. The puzzle is not so much the obvious one that the Lorentz transformation is regarded as a 'precision form' of the Galilean transformation even though in Newtonian mechanics *any* two frames in uniform relative motion are connected by a Galilean transformation. Rather it is that in any consideration involving Newton's laws, Maxwell's equations and constant velocities of observers need to be invoked before the rules for transforming such primitive things as space and time measurements can be formulated. In fact Milne's treatment provides the germ for the required explanation, though its obscure expression has meant that it has not been widely noticed. The set of preferred inertial frames arises directly as a consequence of the commutation requirements which are the conditions for transitivity, and transitivity is the necessary condition for a true equivalence relation. Translating this into our terms now, the existence of the set of preferred inertial frames is simply a consequence of the fact that a number of dcss give equivalent r-times — or, one could say, of there being, in a sense, a universal r-time.

The third point about Milne's treatment can only be referred to here, and a full discussion deferred until later. It is that of the very curious entry of a one-dimensional space in interpreting the results. In his book Milne later "generalizes" the results to three dimensions using an observer's "private

Euclidean space" and this seems an anticlimax after his more fundamental early treatment. We shall find that the three-dimensionality of space enters in a more natural way through the three levels of the hierarchy, so that Milne's difficulty does not arise.

To do better we need in the first place to say something about the interpretation of the external relations between r-times — stating which features of orthodox physics we want to reach. The necessary hint is again provided for us by Milne from the other side of the gap. An external relation between r-times is evidently the feature in the process theory which will correspond in conventional theory to the setting up of a relation between times by sending and receiving light-signals. As Milne is clearly aware but is unable to acknowledge and handle convincingly (see the passage quoted above) the notion of light-signalling has more weight put on it in normal special relativity than it can reasonably bear. We shall therefore now look critically at the usual expositions of special relativity in terms of the (post Milne) k-calculus from our combinatorial point of view. We construct the argument from the point of view of an observer who calls himself B. Another observer, A, passes B sufficiently closely for them to inform each other that they are each in unaccelerated motion at that time. They have machines (clocks) which behave identically when they are together which make ticks and keep counts of them. The counts are called "A's time", "B's time", and so on. After some time B gets a light signal (which may be an association of photons) which has a structure impressed on it (modulated) saying that it came from A and that it set off when A's clock count was t_1. It reaches B when his says t_2. B lets the light signal go on. When B's clock says t_3 it comes back with modulated evidence on it which shows it is the same and that it has been reflected. B lets it go on to A who either reflects it back or sends a new signal with the information that the time then was t_4. We now find that (or else we select experiments so that) if we repeat the experiment we get a constant k in the relations

$$t_4 = kt_3 \text{ and } t_2 = kt_1 . \tag{1}$$

We say that k is a measure of the rate of separation of A and B. So far we follow a standard way of setting the problem (the 'k-theory'). However statements of it are often careless in omitting to stipulate the coding of the signals and leave it to be assumed by default that there is some sort of God-like facility which knows all the times. The k-expounders probably have a vested interest in allowing as much spatio-temporal thinking to be implicitly inferred as possible. When the complete story is told it becomes clear that we are dealing with information transfers in the first place which we should separate from their embedding in familiar language. Indeed our only concession to physics is that we have imported aspects we should call quantum theoretical at one — unextended — point, in using photons or groups of photons or a modulated signal. Once you admit the critique of special relativity (namely that all time relationships must be specified operationally) this seems all you can do. It is true we have used the familiar language about unaccelerated motion and so on to give a clue to what we hope to apply the ideas to later, but really, all observers except B are inferential, and all we can say is that B is receiving a coded signal from some one of them or that he is not. The phrase 'rate of separation', in particular, suggests far more physics than is actually justified by the bare requirements of information transfer. Moreover it would be better not to say that we set up conditions of unaccelerated motion, but that we are interested in conditions when the simple linear expressions for k and the times obtain. If they do not obtain, conditions are simply not those we can handle, though we may look back later, with the theory constructed, and say that the deviation was due to one observer not being in an inertial frame. Similarly we could conduct our argument with an additional constant, a, so that equation (1) would become $t_4 = kt_1 + a$ etc. We should later express this step physically by saying that there had been no need for A and B ever to coincide: they could follow skew lines.

Given one experiment characterized by one value of k we can now write from (1)

$$t_1 t_4 = t_2 t_3 \qquad (2)$$

independently of what value k may have.

Now we reverse the usual k-theory. We can reinterpret k in terms of coordinate length and coordinate time as long as these new quantities satisfy the two conditions (a) that the 'normal' variables must represent k by attaching a number to their ratio, and (b) that this number be of the form $F(x_2, t_2)$.

We now introduce the idea of signals which travel at a velocity c, in order to express the facts about the reflection in terms of lengths and times. This is where what we normally call the assumption that the observers are inertial is made, for the reflection point is inferred. A assigns t', x' to the reflections event as

$$t' = \frac{1}{2}(t_4 + t_1), \qquad x' = \frac{1}{2}c(t_4 - t_1) \qquad (3)$$

because $t_4 - t_1$ is the time of flight there and back and $(t_4 + t_1)/2$ is the time half way along. Similarly for B's coordinates, t'', x'', and these equations become

$$t_4 = t' + x'/c, \quad t_1 = t' - x'/c; \quad t_3 = t'' + x''/c, \quad t_2 = t'' - x''/c \,.$$

Putting these values in (3) and using (1) we get by adding and subtracting

$$t'' = \frac{1}{2}(k + 1/k)t' - \frac{1}{2}(k - 1/k)x'$$
$$x'' = \frac{1}{2}(k + 1/k)x'/c - \frac{1}{2}(k - 1/k)ct' \ . \qquad (4)$$

Write $B = \frac{1}{2}(k + 1/k)$, and define $V = c(k - 1/k)/(k + 1/k)$, and (4) becomes

$$t'' = B(t' - Vx'/c^2), \qquad x'' = B(x' - Vt') \ . \qquad (5)$$

Finally, (2) translates into

$$t'^2 - x'^2/c^2 = t''^2 - x''^2/c^2 \ . \qquad (6)$$

The derived quantity V is the combinatorial equivalent of velocity, and the appropriateness of the definition is shown by putting $x'' = 0$, when

$$x'/t' = c(k^2 - 1)/(k^2 + 1) \ .$$

It may at first sight seem strange that we have emphasized one particular approach to special relativity as being uniquely relevant to our work. To explain this we recur to our starting point. We need to introduce things which vary with respect to one another — length and time being only for the present in the backs of our mind as the final goal. So far, we have had counting, but never *measurement*. We have now taken the first step to introducing it, using an external relation between r-times r_S, r_T where S, T are on different levels. We shall call such r-times *orthogonal*. This involves first an orthogonal relation between two things like the t's which are counts and not yet measurements by themselves. Because we are being combinatorial and not geometric the importance of orthogonality is that we combine the elements of each indifferently rather than stress that they can vary independently, though the latter slant can be a later reinterpretation as a step to getting some geometry. Orthogonal relationships are expected in dimensional structures, and we recall that the hierarchy construction provides a three-dimensional manifold which will automatically be in place in any classification of experimentally generated numbers including k and the values of t.

We are still not out of the wood. To mix the metaphor, there is the ingredient of variability to add in. The construction process of the hierarchy with its ergodic principle (Chapter 6) demands that we select from the unresolved input (in this case the t-numbers in the k-theory) whatever fits the orthogonal criterion. The input will only fit in with this by a lucky chance. In any such construction ('measurement' in the language we have been using) it is k which represents the part which fails to fit with the orthogonal choice in the input and which therefore characterizes the particularity of the experimental situation. In the perfectly ergodic situation experiments have no distinguishable features at all, and for the present argument we provide an underpinning for k by regarding it as a measure of a small deviation from the basic uniform case.

It was necessary for us to give a combinatorial account of relativity in order to get within range of ordinary mechanics, and so to be able to claim to have linked our theory of the origins of the discrete with the language of ordinary physics. However the relativity theorist may well ask whether from his point of view we have done more than recast the special theory in a form that the combinatorially inclined would like. We claim that we have done more. If we take the Minkowski construction seriously then the transformation between time and a space coordinate will be expected to follow the lines of Pythagoras' theorem and this might explain the quadratic form. This explanation is seldom explicitly stated, though most people probably have some such connection in the backs of their minds. This bull has recently been taken by the horns by Pope[3] in his idiosyncratic but provoking 'information based' way of looking at special relativity in which the zero interval measure along light rays is taken literally. Early attempts to stress the symmetry of the time and space coordinates have given way to an attitude in which the Minkowski world is seen as something of a mathematical fancy. If we take the conventional geometrical construction of the Lorentz transformation as self-sufficient then we have to take its fitting in with Pythagoras as a happy accident, which is implausible. Another possibility might be to start from the relativistic equivalence of mass and energy, where, since energy in classical mechanics contains the square of length, we find a quadratic form emerging. The appearance of mass as the time-like component of the energy momentum tensor in general relativity, in which momentum appears squared, makes the same point. To take this equivalence as a starting point would be to make geometry subservient to mechanics. This possibility does not seem to have been taken up.

Attempts to explain the puzzle in special relativity in terms of inertial frames are unsuccessful because the notion of a preferred inertial frame (in Einstein's words "a coordinate-system in which Newton's laws have their usual form") is a more restrictive one in special relativity than it was in Newtonian mechanics. There, any frame could be declared inertial at the expense of admitting certain bizarre fields of force. Hence to use the idea of preferred inertial frame as an explanation is to replace one puzzle by another.

We have removed the puzzle by reversing the argument. The first step is to realize that dcss at different levels require external relations. From this it follows as above that

$$t_2/t_1 = t_4/t_3$$

which is the statement of the constancy of k. Each side is a comparison of different r-times, but the equation is equivalent to

$$t_2 t_3 = t_1 t_4$$

which states an invariance between products of two times of one set and two of the other. The need for the special assumption of orthogonality of coordinates is obviated by the existence of levels in the hierarchy.

The second part of the problem is that the invariant proper-time, which replaces the absolute time of Newton, is a quadratic expression,

$$ds^2 = c^2 dt^2 - dx^2 - dy^2 - dz^2,$$

with the consequence mentioned above that the Lorentz group provides a natural home for the three-dimensional group that the Galilean group did not provide. This second problem, unlike the first, is a purely mathematical one and so has a mathematical solution. The quadratic expression $c^2 dt^2 - dx^2$ is equal to $(cdt - dx)(cdt + dx)$, and $cdt + dx = cdt_4$, $cdt - dx = cdt_1$, so that the occurrence of expressions like

$$c^2 dt^2 - dx^2$$

is an artefact of defining $t = \frac{1}{2}(t_4 + t_1)$, $x = \frac{1}{2}(t_4 - t_1)/c$. This does not go quite so far as to explain the way in which Pythagoras' theorem fits so conveniently. Our explanation of that is again that the usual treatment has been reversed. We get to spatial geometry as a special case of special relativity rather than use spatial geometry as given to set up special relativity.

The second feature of the interpretation of Milne's work which we have just taken over is a more subtle one. From pairs of r-times associated with two dcss at different levels S, T, Milne's method shows how to construct pairs (t, x); new rational numbers. The notation rests on the proven fact that $t^2 - x^2$ is the same for both S and T, and this invariant property corresponds to the roles played by t, x in the Lorentz transformation: t is time-like and x is space-like. To interpret x as a space-coordinate is a bigger step for us than it was for Milne, since he began by assuming space separation between his observers and their clocks. None the less, we shall argue below that it is a correct interpretation.

We start by remarking again the rather curious way Milne begins with a one-dimensional space and then generalizes to three.

In a combinatorial theory "dimension" stands in need of definition. We define dimension using the level structure of the hierarchy. The dimension structure of space corresponds to the three levels. At early stages in the thinking about hierarchies one had partial answers to the question of the origin of dimension. We had replaced the view which is all the theory-language could give that one had length, breadth and height. We concentrated on the symmetry of the dimension structure. This kind of basic symmetry had been analysed in the fifties using the idea of 'similarity of position'. If all mathematical relations remained true however the dimensions were permuted then they were said to have similarity of position (see Chapter 5). This requirement of similarity of position was sufficient to concentrate attention on the quadratic group as the start of the study of dimension structure. One then found an isomorphic structure in the hierarchy level which had strings of two elements (and operators on them of four). It was an obvious step to identify the number of dimensions with the number of discriminately closed subsets at that level, but the identification has turned out to be more complicated and subtle than that.

One has the feeling that the classical macroscopic tripartite division of experience into length, breadth and height is somehow more real than other ways of presenting spatial information. It is probably of little importance to discuss how our human perceptual apparatus happens to work. It would be erring into anthropomorphism to expect that to influence how the world is, and our argument is more general. However for what it is worth one may notice that while there are many ways of using commonsense spatial description which stress that the dimensions are presented all at once, there are many cases where this is not the way in which our preceptions work, and the choice of one way as the real one is arbitrary. Think, for example of the batsman facing a fast bowler. The parameters of the game are adjusted to give him critically little time to think, and he has to bring into action one or other of a large repertoire of strokes and other reactions (including getting himself and his bat out of the way of the ball). He does this on the basis of a fair amount of one-dimensional information — the line of the ball — which is presumed straight until there is reason to think the contrary. If he can get as much as two bits of information about deviations from the straight line (to off or to leg, and how much) he has done well. The third dimension is even more shadowy and relates to the height of the ball above ground. This is by far the greatest uncertainty, and its confusing effects are minimized by moving the bat, which is long and thin, in the plane containing the ball and the batsman's eye. This is called playing a straight bat. More sophisticated variational principles are called into play for more advanced shots: batsmen are not consciously aware of these, but the good batsmen are those who use them unconsciously.

In terms of the hierarchy picture we can say that we have a one-dimensional structure until we find we can't fit everything into that because of the incompatible orderings we find. We need a new structure which we associate with the first. Then the same thing happens and we generate a third dimension. We cannot do it again because of Parker-Rhodes' cut-off (Chapter 5), and we have our abstract three-dimensionality — not to be confused with coordinate spaces. The latter do not arise until we make use of the dual development in the hierarchy of an ordering structure. We may say that the dimension structure handles the unordered information from the background, but it is better to say that the input contains all possible orderings and is subject to the principle of indifferent choice, which has been used throughout the hierarchy construction and in great detail in the calculation of the correction to the coupling constant (Chapter 7).

Our treatment of dimension structure raises problems for discussion. The first is to see how we can accommodate the relativistic notion of time as a coordinate. It is generally accepted by relativity theorists that there is only a conventional, not a real, symmetry between time and the space coordinates. The Minkowski World is a bit of a mathematical fancy. We put the matter more strongly: the 3-D character of space derives from the dimension structure and time is not part of that. In one sense time, as succession, exists before any other structure, because it is one with the process of successive construction, but time is in no sense always there as an ordered manifold, as classical physical time is supposed to be. Given this understanding however, there is no reason we should not adopt the convention that time is a coordinate which occupies the fourth place in the linear space of the hierarchy. This identification will enable us to advance far enough toward classical mechanics to define velocity and acceleration in the usual way. It will also incorporate the relativistic limitations to its own validity at high velocities because that is how it was constructed. The phenomenon on which special relativity depends — the finiteness and invariance of the velocity of light — follows from the existence of a scale-constant. Since we have the coupling constant for electromagnetism which contains the velocity of light as a factor, we can deduce that the finiteness of this constant must exhibit itself as this limiting value. To put it another way; we know that there has to be a breakdown of our classical description in this context, and we regard the particular shape of special relativity as dictated by the need to reconcile this breakdown with a universal continuum dynamics.

A difficulty arises when we concentrate on the *symmetry* of the dimension structure — the emphasis on the orthogonal group, as it were. It has entered the classical theory-language at a very fundamental level, and with very considerable mathematical sophistication, and we have to try to get behind this mathematical veneer. To relate the progression through the levels of the hierarchy with the progressive construction of dimensions seems at first sight to fly in the face of our insistence on the symmetry of the dimensions. Indeed anything which requires us to start with one dimension and then advance recursively seems to offend by giving a preference to one, particularly when the cardinalities of the dimensions are so wildly different. However the point about the symmetry (particularly in the very strong form of similarity of position) is that it must make no difference to the resulting structure which element (discriminately closed subset) we start with.

There is an ancillary argument which supports our use of hierarchy structure in defining dimension. If you have a coordinate system in the sense that you can label points with rational numbers, or the integers, then labelling with three integers is not in reality different from labelling with one: the one case can be said to collapse onto the other. Therefore the construction of dimensions falls to the ground unless it is propped up by something else: in our case the three levels of the hierarchy which — in rather the way Kant said — provide the three levels of experience. Of course if the mathematician is thinking that he has the real numbers up his sleeve then he will not accept this argument, but that we do not allow.

In fact the case for taking the analogy of the batsman facing the fast bowler as a better guide to understanding the perception of space in our process picture than the symmetry of dimensions model lies rather deeper than we suggested. Certainly the hierarchy structure admits exactly three independent external relations to exist 'at one point', but we do not take the expected next step of identifying the three external relations with three (orthogonal) dimensional directions. The different complexity of the different levels is not the most serious reason for this. More serious is the failure to provide any description of what lies "in between". There is a place in it for North and East and Upwards, but none for a direction NNE at an elevation 5° as it were. Such questions do not arise for us.

We now discuss some possible criticisms of our approach to conventional spatial ideas. The change we propose is so surprising to the normal thinking that we prefer to look at the first steps from many points of view so as to make the need for them clear rather than to get into much technical detail.

The view we have taken, that we can establish the first fundamental steps and then fill the picture of space in progressively, constitutes the first difficulty — perhaps the most profound — that readers may find. One naturally slips into thinking that if once we can make any statement about a spatial relation we have in principle the whole of space at our disposal. It is therefore puzzling in itself that we deal with spatial constructions one by one — making a careful start, but being prepared to leave even the work of reconstructing the classical dynamics of a particle to a later stage. We have insisted that our theory is one of process, or that it is constructive: piecemeal build-up of the dynamical concepts is a consequence. On the one hand we have to explain some basic things which are assumed without explanation in both classical physics (and more culpably in quantum physics) like why particles travel in straight lines with an impression of continuity, and on the other hand we abandon the notion that when we have constructed some dynamical situation a surrounding universe is in principle automatically implied, and could be described by simple extension of our method.

We now consider a different objection which requires us to be very explicit about applying our ideas of constructivity to space in order to answer it. When we seek to discuss such a thing as Pythagoras' theorem in a discrete space, we need to represent the spatial difference between two points a, b as a difference $a - b$ of coordinates in say, one dimension. Further, if you have three points in a line then you are committed to saying that the distance ab is the sum of ac, cb. To see the problem which arises in a discrete space, choose units to give the line unit length (1). We have either a finite or a denumerably infinite number of points. So we enumerate them. The way of doing this may be quite subtle, but, one way or another there are first, second, and so on points. Now enclose one new point (1) with a little segment of length $1/4$, point 2 with a segment of length $1/8$, and so on. We know that the total length of the segments will not exceed $1/2$. Yet the line has length 1, and we have enumerated all the points. It seems to make no sense to say that the line has length 1, let alone that we have enough apparatus to formulate Pythagoras' theorem.

To avoid this conclusion we have to argue that the argument which we have just given is not finite and constructive. One cannot talk about metric distance without constructing a total ordering; putting the labels in an agreed way; and agreeing to a rule to represent ordinality. The alleged contradiction involves a circularity because the notion of a covering piece for each point already assumes the definition of distance that is being questioned. McGoveran takes finiteness and constructivity as axioms in his *Ordering Operator Calculus* whereas the authors argue from physical necessity. McGoveran is accordingly able to be much more definite than we when he insists that spatial properties have to be constructed with each special application clearly in view, and our thinking owes much to him.

Earlier in our combinatorial work we simply asserted that the particular shape taken by relativistic mechanics was not of great significance provided only that it incorporated the bounding character of the velocity of light, because that was dictated by the structure of scale constants. We stand by that position so far as it went, but have now improved it by establishing an entree into special relativity and hence into the physical representation of space and coordinate time. Further we have not gone. Newton's laws are obviously needed. It may seem a funny idea to expect to explain them since in the usual physics they are just laws, or you might say, axioms.

* * * * *

Let us recap. What do we need for classical space? We have dimensionality, but not in its familiar form where it is inextricably connected with the idea of a coordinate system, which we have not got. We certainly think it an advance to have separated the aspects of dimensionality out in the way we have, and yet the symmetry of the coordinates should be part of the total picture. We can only plead that to have the coordinate structure with its full meaning is further than we are able to go at present. Our problem is to relate the 3D from level structure to the 4-vectors in the labelling system. As appears in detail from the hierarchy mathematics the labelling scheme has to be described by a linear algebra of order 4, 16, or 256. It is straightforward to see the higher level algebras as descriptions of complex structures in the simplest, rather as tensors of ranks 2 and 4 appear in general relativity. However we have to justify regarding the fourfold structure of the labelling system as a coordinate system which takes over the three-dimensionality from the level structure (quite apart from that fact that the label strings are over Z_2 and bear only a structural resemblance to a coordinate system). To reach ordinary commonsense physical space we need the recursive memory structure potentially available from the hierarchy algebra to be exploited, and this has not yet been done. An entree has been established by relating the symmetry of the coordinates to the independence of the hierarchy levels, and noticing that this connection provides understanding of the geometrical notion of orthogonality. Given this understanding it is a natural step to derive from the individual invariants $T_S^2 - X_S^2$ etc. a combined invariant $T^2 - X_S^2 - X_T^2 - X_U^2$ where T^2 is $T_S^2 + T_T^2 + T_U^2$. Thus, rather surprisingly the natural route in our theory is through the connection between space and time in special relativity to a conclusion that Euclidean geometry holds, rather than in the opposite direction.

CHAPTER 10

Objectivity and Subjectivity — Some 'isms'

1. Subjectivism

The quantum theory has subjective aspects. This is denied by some physicists (like Peierls) who argue that 'measurement' is a scientific phenomenon as much as those on which pre-quantum physics was based. Our earlier criticism of this use of 'measurement' (or 'observation') can be summed up in the question: "In that case why call it measurement?" If it were given some neutral name then the tricks it is used to cover up would no longer seem plausible. Our theory is not subjective in the same way because it does not introduce a special class of fundamental physical processes associated with human activity. However it is about sequential processes through which knowledge of a background world can be obtained, and therefore puts a cognitive element in the most elementary forms of being, so that that cognitive element does not have to be imposed upon an alien conceptual framework as happens in current quantum theory. The knowledge needed to construct basic physics is very primitive and does not presuppose anything like a conscious observer, so that we can imagine it going on without there being physicists. On the other hand there is no reason why conscious observers should not participate in the process, and this is our resolution of the observer paradox.

2. Realism

Older people can remember living through a great conceptual change in their understanding of time. At the beginning of this century no one doubted that of every two non-simultaneous events there was one which came first and one which followed, even though it may not have been possible to discover the order experimentally. Now, we really have abandoned that preconception and really feel in our bones that the ordering of the events depends on the means available to determine that order. The theory presented in this book requires a similar change in the understanding of time, and one that follows on naturally from the first. What now has to be abandoned is the assumption that something unique must have happened at every previous stage in the development of the universe whether we happen to be able to discover it or not. The information — so the preconception has it — is there and potentially recoverable whether we are able to dig it out or not. For us, time specifies relations between objects just as space does, and what we know of objects resolves into the sequences which we have been studying if we push the analysis far enough. We have already located the origins of this relational view of time and space in the writings of Leibniz.

At an everyday level the difference between the views may not obtrude. We can say that the history of the Tudor period is a flow of time in which such-and-such events took place, or we can say that we have records of events which are so related as to be conveniently spoken of in a certain sequence which we call 'time', and which form of words we use doesn't much matter. When we come to cosmical times things change. It is really misleading to speak of a beginning in time (for example in the context of the big bang theories) because the word 'time' is really only appropriate when there are many cross-referencing ways of specifying relations between events. In the case of the Tudors, we can go to the library, or we can do archæological digs, or many other things. In the cosmological case, the further we extrapolate, the fewer become the independent lines of experimentation, and the more these lines become identified with the structure which generates the scale-constants. Wittgenstein's aphorism "The limits of my language are the limits of my world" seems to apply with a literalness he would probably not have countenanced. At the limits of extrapolation (in time or in space; on the large scale or at the small) we are brought face to face with what are really the conditions for measurement to be possible, but which masquerade as ordinary results of measurement.

3. The anthropic principle

In recent years the place of mind in the universe has been discussed from a new standpoint. Minds can only survey universes which are suitable for their existence, and our judgments about the plausibility of particular theories ought, it is argued, to recognize that a sort of selection procedure has taken place out of all that might have been, and if possible to allow for its effects.

Many strands of argument concerning the anthropic principle have been analysed in detail, particularly in the book of that name by Barrow and Tipler.[1] Of these, the strand which comes nearest to numerical precision concerns the coupling constants — what we have called the scale-constants. It is commonly accepted that if these constants did not have the values they have, or values remarkably near, then the universe would be very different, and we cannot be at all sure that it would support life of the sort we know. In that case the anthropic principle would lead us to think of a range of universes each characterized by a particular set of numerical values for the scale-constants — our particular universe being preferred only because its set of values enables us to be here to know about it. A corollary of this view is that it is a mistaken endeavour to try to explain the values of the coupling constants.

An alternative way to put the argument is to start from the position that the scale-constants are simply sliding parameters (particularly as they have shown such recalcitrance to being understood or calculated) in which case the anthropic principle removes the puzzle of why the conditions needed for intelligent life are so surprisingly exactly catered for.

One may suspect these arguments on two general grounds. Both come from the suspicion that in circumstances in which fundamental changes were being imagined in the foundations of physics, too much may have been taken for granted about the way possible worlds and their occupants would still resemble our familiar one. In the first place, one does not have to take the radical views put forward in this book about the nature of the scale-constants to be wary of treating them as disposable parameters. We should be wary anyway. It is accepted in current physics that they are very fundamental, and in the absence of any understanding of their origin it is only speculation that they could be changed without making the experimental basis of quantum theory unrecognizable. This point leads to the second ground for suspicion. Do we really know how much life forms are bound by what we know of physical theory? One slips back into thinking that matter is explained by quantum mechanics and life-forms are made of matter, but even from a quite orthodox standpoint one has to be very critical of such unconscious extrapolation. No one has deduced Newton's laws (even the straight-line motion of a particle in the absence of constraints) from quantum mechanics. Much less ought it to be claimed that the complexities of biophysics follow from a quantum-mechanical foundation, though people assume they do.

Now if we allow, for the sake of argument, the claims about the scale-constants made in this book, we immediately get a clearer perspective. If the scale-constants are derived at a more primitive stage of theory construction than the rest of quantum theory and are in fact necessary conditions for any theory, then it makes no more sense to raise questions about the effects on the probability of conscious life emerging if the constants had different values, than it would to ask the same questions about the effects of changes in the value of π. In short, no sense at all; and inevitably that is our position on the anthropic principle.

4. Constructivism

If we try to link our way of developing the mathematics we use from the idea of process, with some acknowledged position on the foundations of mathematics, then we come nearest to 'constructivism'. Indeed in his treatment, McGoveran prefers to start with an axiom requiring constructivity to starting with ideas of process as we have done. Constructivism was the logical innovation of L. E. J. Brouwer. Brouwer's demand that mathematical entities be constructed by defined steps was a restriction on the freedom of the mathematician which needed to be compensated by his vision of a mathematical universe of 'freely proceeding sequences'. Brouwer told one of the writers (T.B.) that he was quite happy with the idea that the empirical data of physics might dictate the progress of the freely proceeding sequences. He did not see this as contradictory with the appropriate kind of freedom.

5. Reductionism

Reductionism is the doctrine that everything about the world follows in principle from the set of laws which govern the behaviour of elementary systems — particles and/or fields. Thus we get, according to this doctrine, a scale of increasing complexity starting with physics and usually going on through chemistry and finally reaching the biological organisms. In this doctrine, however, there is scope for additions to be made to theory in the light of later knowledge. For example it would not be held as against the spirit of reductionism if principles from information theory or cybernetics were found to be necessary to bridge gaps between different stages of explanation. Moreover there are two forms of the doctrine which we may call the strict form and the weak form respectively. In the first, everything is, at least in principle, deducible; in the second, new properties are allowed to emerge as the complexity grows.

Combinatorial physics is incompatible with reductionism in both of its forms, because the atomic structures it provides are not logical atoms at all. They are the forms in which the unknown things about the universe are bound to present themselves, but they are not themselves the unknown things. They are not the building bricks from which everything is constructed.

It is worth making this clear disavowal since combinatorial physics does not fall neatly into any of the categorizations of theories. Some will feel antipathetic towards it on the ground that it leaves the way open to a mystical apprehension of the world. It probably does, though our only criteria in judging its plausibility as it has developed have been internal to physics: it must make its way as physics. In any case the material realism which most physicists probably feel to be the natural scientific attitude is usually set against a very crude theism, (Dawkins[2] *The Blind Watchmaker*, is a good example of this crudity) whereas our efforts would not afford support for either side. Our position is no more tolerant of crude theism than it is of material atomism.

6. The critical philosophy

We should be happy if this book were to re-open discussion about which of the traditional philosophers best represents the physics of the quantum. Some of the founders of the quantum theory thought that Kant got nearest. We recall that in our mathematics we make the important decision to treat all the choices of constructed elements that are allowed by the mathematics indifferently. If we do not do that we do not get our numbers right. However we are making an assumption about the unknown background which is not forced on us. Indeed, what we could call truly empirical knowledge will always come through some sort of weighting of the input — the order in which the unknown elements present themselves. That is our only access to the unknown. These therefore are like the Kantian noumenon, the Ding an sich. However when we look at space and time, which for Kant are the *a priori* necessary forms of intuition, we find we have changed from his position. Space, at least, is constructed explicitly by us, and is therefore dependent on the indifference assumption. It is not *a priori* necessary, and our thinking encourages us to look for other ordering principles if we want (as some may wish to do) to apply constructive thinking to a wider class of evolutionary situations. So our view is Kantian only up to a point, though it seems that from our point of view Kant was righter than the more modern philosophers.

7. Positivism

The reader may well ask how philosophers and philosophers of science react to the philosophical aspect of combinatorial physics. The answer seems to be that for the most part they would not consider it at all because they do not take the philosophical problems of quantum physics as a whole seriously. We recall how the founders of the quantum theory were driven to postulate an epistemological foundation for quantum theory which was heavily out of keeping with consensus philosophical opinion. The fact that we have criticized current quantum theory for its failure to present the epistemological intrusion coherently in no way means that we are content to retreat from the facts and the thinking that caused the founders of the theory to think it vital. We think the hedges have to be jumped, not refused. This, however, is not the view of the philosophers, who want neither change in the metaphysics of science nor accept that modern physics has anything new to tell us about mind.

It has to be admitted that the encapsulation of the new epistemology into one mathematical device (say, the wave-function collapse) which, once swallowed, enables the physicists to resume business as usual, plays into the hands of the philosophers. They (who of all people ought to kick at the device instead of swallowing it) are encouraged to treat the device as a part of the technical development of the theory which they accept as given. The philosophers of science in particular devote much time to detailed glosses on detailed points of the theory on the basis that the theory as a whole has to be accepted, and all they can ever do is to find more acceptable ways of putting the regrettably wild things the physicists say.

The extensive writings of Popper on the principles of the quantum theory, fall into the class we describe. Popper is broadly opposed to the Copenhagen interpretation of the theory. He therefore has the problem of improving on the story that measurement intervenes in a physical process to cause what is sometimes called the collapse of the wave-function. He retains the mathematical structure which gave rise to that description and turns his attention to probability to explain the strange way the wave-function behaves. Wave-functions are probability distributions, and collapsed wave-functions are a special case in which the probability of one result is nearly unity and that of all the rest nearly zero. The implausibility of the wave-function collapse is that one can see no reason for such a change in the distribution. It seems very improbable on any dynamical ground. Popper[3] introduces a *propensity* view of probability, according to which things have a propensity to behave the way they do. If they choose to collapse; well, it's a free country.

We are not concerned with the rights and wrongs of the propensity account of probability here nor with whether it makes the wave-function collapse more comprehensible. The important question stands unchanged, it is: Why do some preferred physical processes have this collapsing kind of propensity and not others?

Our problem is now clear. We want to present to the philosophers an improvement on the mainstream-quantum-physics view that mental process at the human level of complexity is needed to make physics work. We want to claim that if one is forced to discuss mentality (consciousness if you prefer) then one must have it at every stage of the physical constructive process right from the primaeval decision process with two-bit strings. We might have expected that this reinstatement of an age-old philosophical issue into a state of active enquiry would be seized upon to widen the boundaries of the sadly narrowing discipline of philosophy. What, in contrast, we are likely to find in most (perhaps not all) quarters is that the repudiation of the physicists' metaphysics of quantum physics (as being known beforehand to be wrongheaded and confused) which we have just described, will pursue us too, whether our case is more cogent than that of current quantum theory or not.

Here is a real puzzle. We see well enough that the positivistic criteria were needed to sweep away metaphysics which had become detached from its origins in experience, but what can possibly make the philosophers so sure that any metaphysical questions raised by the active promulgation of science are to be ruled out of order automatically? The answer to this puzzle that we have been forced to is rather shocking, and likely to seem arrogant. *However we can no other*: it is this. The philosophers have an even more dogmatic adherence to the commonsense Newtonian picture of the world than the average scientist has, and unlike him they allow themselves to remain ignorant of the unsettling discoveries of this century by regarding them as technicalities which are not their business. As a result 'their business' has become progressively restricted to a few (admittedly deep) issues on the fringes of logic, and the opportunity to breathe new life into some long dead philosophical concepts is being missed. In particular we are still left with the ossified Cartesian matter/mind distinction. We know what matter is, and 'mind' is just a funny word with no particular place in intelligent discourse.

8. Operationalism

Operationalism is the deepest and most abiding philosophical theme of modern physics. It is a leaven which works slowly but leaves, at last, nothing unchanged. It is usual to trace its roots to Mach. In pursuit of his view that all knowledge is derived from "sensations", Mach took the first step toward what we should now call operationalism by saying that phenomena under scientific investigation can be understood only in terms of experiences or "sensations", present in the observation of the phenomena. This is a form of the verification principle, but is more specific about what constitutes verification. Mach's criteria led him to reject such concepts as absolute time and space, and had the result that the Einstein relativity theory was less of a shock than it might have been. The picture at the quantum scale is less clear. Mach's operational thinking evidently led him to deny the reality of the quantum objects on the grounds that they were merely artificial constructs from sensations. While in this book we have found it necessary to question the presumptions of simple realism to the particles, and in that sense go to some extent with Mach, we are far from denying them any reality. Our picture is far more complex and far more interesting.

Bridgman gave the approach its modern and tightest form by requiring that verification be referred to well defined experimental operations or sequences of them. In the central case of relativity, for example, the primary operation uses a ray of light. What a ray of light will not do, as in establishing simultaneity, cannot be done. Bridgman's philosophical position was extremely radical: a concept meant the procedures used in its observation. "In general, we mean by any concept nothing more than a set of operations; the concept is synonymous with the corresponding set of operations. If a specific question has meaning, it must be possible to find operations by which answers may be given to it. It will be found in many cases that the operations cannot exist, and the question therefore has no meaning." (Logic of Modern Physics.)

Bridgman's remark about operations which cannot exist rings alarm bells. Can he be taken literally? How could we possibly know that they could not exist except through some theoretical understanding to which we are allowed at this stage no recourse? The paradigm case of the operational critique is to be found in the history of relativity, where the vital experimental operation was the use of light rays. However the consequent sacrifice of absolute time was all part of a general theoretical understanding which made sense of the Michelson–Morley experiment. This history suggests that the understanding of the operational position and of the theory are likely to advance hand-in-hand rather than that the operationalist would stand outside the developing theory to confer or withhold his unchanging *nihil obstat*. Our unease is increased because Bridgman sometimes watered down his operational critique by allowing that operations that could be imagined would count. One understands that in, for example, cosmology, many statements are made which Bridgman would have wished to admit as meaningful but which could not in practice be tested. It does not help to say that theory gives no reason why they should not be tested in principle. The power of the operational criterion lies in the demand that one answer the question, "Can you actually do the operations or can't you." That was the cutting edge in the early days of relativity.

There seems no way out of the conclusion that a lot of ideas we actually work with at present do not conform to the operationalist critique. However in our view it would be a mistake either to impose an unrealistic operationalist fundamentalism on the one hand, or, on the other, to ignore the critique. Perhaps one should rather treat it as a signpost or a warning.

There is one wide area over which the operationalist warning seems universally to be ignored in favour of the current commonsense, and that is in extrapolation over cosmical distances and times. Bridgman accepted it as a consequence of his position that the concept of length, for example, on the laboratory scale (metre sticks and so on) was a different concept from the thing of the same name on the cosmical scale (the red shift interpretation dominating the technique of operation). It will be a long time before Bridgman's realization is taken literally, for it is deep in the heart of the scientist not to question the assumption that there is a simple spatial reality which unifies the length concepts, and one which it is quite unnecessary to investigate. Even as we write, evidence from the Hubble photographs show a discrepancy between red shift scales of time or distance and those from cepheid variable frequency counts. All sorts of ways of reconciliation are being proposed but no one has suggested that if the assumption of an underlying space concept which unifies the various experimental operations were to be looked at critically then quite new perspectives (as radical as those liberated by the critique of simultaneity at the beginning of this century) might appear. The first impact of the new freedoms might seem negative at first, but even the initial clearing of the ground would be reassuring. Many people must have worried about the wild extrapolations of meaning that accompany discussions of the Big Bang, because of this lack of awareness of the dangers in changing the meaning of concepts as their operational status changes.

The operational position on space in quantum physics has concerned us in this book at an even more basic level than that on space in relativity. Where the particle attributes originate, space does not exist at all. However, the gap between what we are saying and current thought is not as great in quantum physics as it is in relativity. Thirty years ago Chew[4] ("The dubious role of the space-time continuum in subatomic physics") had this to say. "My thesis in this lecture is to suggest that the concept of space and time is playing a role in current sub-atomic physics analogous to that of the ether for macroscopic physics in the late nineteenth century physics. It may never be possible to demonstrate that a space–time continuum cannot exist, but a growing number of us are concluding that, to make major progress, we must stop thinking and talking about such an unobservable continuum." In reminding us of what happened to the ether, Chew is as good as making an explicit appeal to the operationalist critique.

Of course we have an axe to grind. In a theory of process, all statements about spatial relationships have to come with the ways of constructing them specified. This requirement is as rigorous as the need to specify simultaneity which everyone now accepts, and in this book we have taken the initial steps in formulating relativity in such a way as to make spatial relationships constructive in the same way as temporal ones.

9. Particles

At the end of the day, what are the particles? They are not matter cut up as small as possible — Newton's small hard massy particles, 'even so very hard as never to break or wear away'. They are not convenient fictions, as the Nineteenth Century field theorists would maintain, either. The quantum theorists claim to have a mechanics which has had some successes in seeing them as self-generating, and with properties which have a logical necessity (think of exchange forces and principles of exclusion). However their theory leaves unanswered the question of the nature of the thing to which these theoretical ideas are attributed, and by default that thing comes to be thought of as a spatially located commonsense object of a thoroughly Newtonian sort. The answer we give, by contrast, is that the particles are fully objective and real in some senses of those words, but not in others. Physicists are to a dangerous degree accustomed to filling out a little universe of conventions round the little bit of information that we really do have about some particle. In classical physics this is always justified because of the nature of our experimental knowledge; in quantum physics it is not, but the quantum theory provides no conceptual framework which enables us to move comfortably in the new situation. In the theory we have proposed, the sliding scale of available information with the quantum objects at the bottom is a central theoretical idea. It is all-of-a-piece with the mathematics which has given new calculations about the particles, and therefore it is only a matter of getting familiar with it.

The condition for our having a spatio-temporal framework at all is to understand the interacting processes from which we have been able to derive it, and these interactions have numerical limits. In a way we are debunking quantum physics. Atomic objects seem to have incredible properties (we could call them 'logical' properties to contrast their sharpness with the fuzziness attendant on all classical objects if we look at them closely enough) but the incredibility comes because the logical structure which they seem to exhibit in a pure form is wrapped by our thought habits in the trappings of conventional objects. In fact these 'logical' properties come as the condition for observation to be possible.

To say this does not entail the complete subjectivism of saying that the properties of the particles can be deduced from a study of how we measure them, as some have argued. Since the conditions of measurement are under our control, and can be altered at will, the properties of the particles must, on that view, depend on our human decisions. For us, by contrast, these properties are as real as anything else. They differ because the information they carry comes from somewhere else where the particle picture will not penetrate and which we need to think about with ideas appropriate to whole sequences rather than with concrete images describing their separate steps.

References

Chapter 1
1. H.J. Ryser, *Combinatorial Mathematics*, Math. Assos. America, Wiley, 1963.
2. G. Polya, *Introduction to Applied Combinatorial Mathematics*, ed. Reckenbach, Wiley, 1964.
3. H. Weyl, *Philosophy of Mathematics and Natural Science*, Princeton, 1949.

Chapter 2
1. "Concept of order I", *Proc. Camb. Phil. Soc.* **50**, 2 (1954) 278.
2. A.N. Whitehead, *Adventures of Ideas*, Cambridge, 1947, p. 168.
3. B. Russell, *History of Western Philosophy*, Allen and Unwin, 1946, 2nd ed., p. 576.
4. L. Couturat, "Recent work on the philosophy of Leibniz", *Leibniz*, ed. Frankfurt, Notre Dame Press, reprint 1976.

Chapter 3
1. M. Born, *Atomic Physics*, Blackie, 3rd. ed., 1944, p. 144.
2. Address at the International Congress of Anthropological and Ethnological Sciences in Copenhagen, delivered at a meeting in Kronberg Castle, Elsinore, August 1938. This essay appeared in *Nature* **143** (1939) 268, and was reprinted in Bohr's book *Atomic Physics and Human Knowledge*, Wiley, 1957.
3. *Atomic Theory and the Description of Nature*, Cambridge, 1934, p. 52.
4. A.F. Petersen, *Quantum Physics and the Philosophical Tradition*, M.I.T, 1966.
5. *Symposium on Nature, Cognition and System*, 4th International Conf. on Systems Research, Baden-Baden, August 1988.
6. See the discussion by B.J. Hiley, "Vacuum or Holomovement", *The Philosophy of Vacuum*, ed. Simon Saunders and Harvey R. Brown, Clarendon Press, Oxford, 1991.

Chapter 4
1. Y. Aharonov and A. Petersen, "Definability and measureability in quantum theory", contribution to *Quantum Theory and Beyond*, ed. Bastin, Cambridge 1971.

2. N.R. Campbell, *Physics, the Elements*, Cambridge, 1922.
3. A.S. Eddington, *The Relativity Theory of Protons and Electrons*, Cambridge, 1936, and *Fundamental Theory*, Cambridge, 1946.

Chapter 5
1. A. F. Parker-Rhodes, Earliest reports of his combinatorial work are in "On the origin of the scale-constants of physics", Ted Bastin, *Stud. Phil. Gandensia*, 1966, p .77, and "On the physical interpretation and the mathematical structure of the combinatorial hierarchy", (Bastin, Noyes, Amson, Kilmister), *Int. Journ. Theor. Phys.* **18**, 7 (1979) 445. His book, *Theory of Indistinguishables*, Reidel, appeared in 1981.
2. E.W. Bastin and C.W. Kilmister, *Concept of Order I, Space-Time Structure*.
3. J. Conway, *On Numbers and Games*, London Mathematical Society, 1976.

Chapter 6
1. A. Watson, *The genesis of structure: a twentieth century Copernican revolution*, University of Sussex Ph.D. thesis, 1991.

Chapter 7
1. H.P. Noyes and D.O. McGoveran, "An essay on discrete foundations for physics", *Physics Essays* **2**, 1 (1989) 76.
2. E.W. Bastin, *Stud. Philos. Gandensia* **4** (1966) 77.
3. R.P. Feynman, *Theory of Fundamental Processes*, Benjamin, New York, 1961, p. 33.
4. Wyler, *Comp. Rend. Acad. Sci.* (Paris) **269A** (1969) 743.
5. D.O. McGoveran, Proceedings of ANPA 9, published ANPA, c/o F. Abdullah, 1989. The exposition is based on McGoveran's ordering operator calculus which is to be found in D. McGoveran and H.P. Noyes "Foundations for a discrete physics" in *Physics Essays* **2** (1989) 76–100.

Chapter 8
1. H. Pierre Noyes and David O. McGoveran, "An essay on discrete foundations for physics", *Physics Essays* **2**, 1 (1989).

Chapter 9
1. E.A. Milne, *Kinematic Relativity*, Oxford, 1948.
2. E.A. Milne, *Zeitschrift Astroph.* **15** (1938) 270.
3. N.V. Pope and A.D. Osborne, "Instantaneous relativistic action at a distance", *Physics Essays* **5**, 3 (1992) 409.
4. G.J. Whitrow, *Quart. J. Math.* (Oxford) **6** (1935) 249.

Chapter 10
1. J.D. Barrow and F.J. Tipler, *The Anthropic Cosmological Principle*, Oxford, 1986.
2. R. Dawkins, *The Blind Watchmaker*, Longman, London, 1986.
3. K. Popper, "Quantum theory and the schism in physics", ed. W.W. Bartley III,

Hutchinson, 1982. The discussion (p. 205) of the propensity view in the context of aspects of traditional philosophy is particularly interesting.
4. G. Chew, Lawrence Radiation Laboratory, Berkeley, Pub., Rouse-Ball lecture, Cambridge, 1963.

Name Index

Aharanov, Y., 26
Amson, J., 38
Amson invariance, 127
Aristotle, 13

Barrow, J and Tipler, F., 155
Bell's theorem, 26
Berkeley, 49
Bohm, D., 27
Bohr, N., chapter 3, 26, 27, 49, 96
Bohr atom, 96
Bohr-Sommerfeld treatment, 96
Bondi, H., 140
Brouwer, L.E.J., 43, 63, 156
Brouwer's intuitionism, 43

Campbell, N.R., 28
Cantor, G., 7
Church's thesis, 50
Conway, J., 86
Conway's trick, 85
Couturat, L., 15

Dawkins, R., 157
Dedekind, R., 7
Descartes, R., 14

Descartes' vortices, 10
Dirac, P.A.M., 17

Eddington, A.S., 30–32, 93, 124
Eddington's conjecture, 30, 31
Einstein, A., 26–27, 145

Fermi (4-particle) coupling, 93
Feynman, R.P., 93, 123

Heisenberg, W., 18, 26, 123
Higgs, P., 130
Higgs boson, 130
Hilbert space, 137

Jeans' "Report on radiation and the quantum theory", 25

Kant, 21, 149, 157

Lamb shift, 137
Leibniz, 2, 9–15, 59, 105, 154
Lorentz contraction, transformation, 11, chapter 9
Lucretian doctrine of the void, 12

McGoveran, D., 91–101, 151, 156
Maxwell equations, 94
Milne, E.A., chapter 9
Minkowski world, 145, 148

Newton, 8, 9, 12, 13, 15, 105, 124, 125, 145, 146, 151, 155
Newton's small hard massy particles, 162
Noyes, P., 41, 42, 81, 91, 92, 124–132

Parker-Rhodes, A.F., chapters 5, 6
Parker-Rhodes bounds, 79
Pauli exclusion principle, 19
Peierls, R., 17
Petersen, A.F., 21, 26, 27, 49
Poincaré, H., 32
Poisson bracket, 137
Polya, G., 2

Pope, V., 145
Popper, K., 158
Planck, M., 20, 25, 30, 129
Pythagoras' theorem, 145, 146, 150

Russell, B., 14, 15
Ryser, H. J., 2

Sommerfeld, A., 92

Thomson and Tait, 8
Turing machine, 50

Watson, A., 58
von Weizsaecker, C.F., 44
Weyl, H., 3
Whitehead, A.N., 12, 13
Wick, G.C., 122
Wittgenstein, L., 9, 15, 154
Wyler, 94, 99

Yukawa, H., 122, 123
Yukawa (3-particle) coupling, 93

Subject Index

a priori, 21, 31
a priori necessary forms of intuition, 157
abelian group, 34, 68
absolute scale, 29, 30
addition mod 2, 34, 46
address, 50, 54
Amson invariance, 127
anthropic principle, 155, 156
array, 30, 78–89, 127
Ars Combinatoria, 3, 15
attribute d-space, 95

background universe, 59
baryon number, 130
batsman, 147, 148
Bell's theorem, 26
binary string, 35
bit-string pattern, 96
Bohr atom, 96
Bohr–Sommerfeld treatment, 96
broken symmetry, 130
Brouwer's intuitionism, 43
bubble chamber, 122

cartesian matter/mind distinction, 159
characteristic function, 72–80, 113

charge, 30, 122, 130–133
Church's thesis, 50
combinatorial, chapters 1, 2, 3, 4, 5, 6, 7, 9, 10
combinatorial event, 96
combinatorial hierarchy, 34, 59, 91, 95
combinatorial physics, 12, 15, 157, 158, chapter 4
complementarity, 3, 17–24
completeness, 26
concurrent computing, 10
congruence, 139
conjugate variables, 101
conservation in processes, 130
constructivism, 156
content, 55
content string, 129
context-independence, 59
continuum, v, 7, 122, chapters 2, 4, 9,
continuum physics, 25, 29, 135
Conway's trick, 85
Copenhagen interpretation, 158
correspondence principle, 13
cosmical time, 154
counting, 27–32, 41, 48, 91–93, 144
coulomb interaction, 97, 130
counter-firing, 41
coupling constant, 37, 92, 93, 121, 148, 155
critical philosophy, 157
crude theism, 157
cumulative sum, 37, 53
cybernetics, 156
cyclic group, 34, 87
cyclicity, 96, 97, 99

Descartes' vortices, 10
descriptor, 43, 47, 122, 123
differential operator, 3, 137
dimension, 31, chapters 5, 6, 7, 8, 9
dimensionless constants, v, 29
Ding an sich, 157

discriminately closed subset, or dcs, 35, 147, 149, chapters 5, 6
discrimination, 34, 109, 126, 129, 133, chapters 5, 6,
discrimination system, 38, chapters 5, 6

Eddington's conjecture, 30, 31
eigenstate, 3, 137
Einstein–Podolsky–Rosen Gedankenexperiment, 26
element, 2, 30, 31, chapters 5, 6
entity, 20, 43, 57–83, 101, 125
epistemological intrusion, 158
equal, 46–55
equivalence class, 28, 66–68
Euclidean geometry, 152
exchange, 122–124, 130, 162
exclusion, 3, 19, 137, 162
exhaustible set of attributes, 55
exoskeleton, 52
extrapolation, 9, 11, 12, 131, 136, 154, 155, 161

Fermi (4-particle) coupling, 93
fine splitting, 92
fine-structure constant, v, 31, 92, chapter 7
finite field, 68
freely proceeding sequence, 156
function, 45, 65–90

Galilean transformation, 141

hierarchical model, 35, 57, 131
Higgs boson, 130
high energy scattering, 41
Hilbert space, 137

identity of indiscernables, 59
incompatibility, 21–23
indistinguishables, 59, 76
infinite tail, 77, 106, 107, 127
information theory, 156
integral/half-integral spin, 129
interaction logic, 49

internal and external procedures, 103
irretrievable, 125
isospin, 128, 133

Jeans' "Report on radiation and the quantum theory", 25

k-calculus, 136, chapter 9

label, 37, 39, 42 et seq., 45–55, 99–105, 124–133, 137, 149, chapter 6,
Lamb shift, 137
lepton, 123
level, 36, 125–133, chapters 5, 6, 7, 9
light-signalling, 136
limited recall, 45, 50
linear characteristic function, 77
linear equivalence, 103, 140
Lorentz contraction, 11
Lorentz transformation, 11, chapter 9
Lucretian doctrine of the void, 12

mapping space, 96 et seq., 125
mass, 8, 19, 29, 93, 123, 131, 145
material atomism, 157
matrix numerical method, 116, 117
Maxwell equations, 94
memory, 42, 54, 61, 68, 151
metaphysics, 15, 49, 58, 158, 159
mind, vi, 154, 159
Minkowski world, 145, 148
monad, 12–14, 48

naive realist, 21
Newton's small hard massy particles, 162
non-commutativity, 137
noumenon, 157
nym-addition, 46

objectivity, 14, 153
observation logic, 45, 49
Old Quantum Theory, 96

operationalism, 129, 159
ordering operator, 91, 94, 98, 151
ordinal scale, 28
orthogonal, 95, 144–152

parity conservation, 131
Parker-Rhodes bounds, 79
particle (classical), 20
particle (quantum), 20
Pauli exclusion principle, 19
periodic, 96
perm, 38–41
PICK, 54
platonic receptacle, 9, 135, 136
plenitude, 14
Poisson bracket, 137
positivism, 5, 154
preferred inertial frame, 136, 141, 145
preparation of states, 41
primary mass ratio, 123
principle of choice, 62, 74
probability amplitudes, 121
process, 1, 9, 34, 57, 58, 65, 69, 126, 135, 136, 142, 150, 156, 162
process universe, 57
Program Universe, 54, 126, 129
propensity view of probability, 158
proton/neutron, 128
Pythagoras' theorem, 145, 146, 150

quantum field theory, 26, 122
quantum number, 33, 43–55, 91, 96, chapter 8
quantum postulate, 20, 23, 41, 49
quark, 123, 130

r-time, chapters 7, 9
ratio scale, 28
real numbers, 7, 149
realism, 22, 124, 153, 157, 160
realist 'commonsense', 124

recursion, 42, 43, 76, 85, 124
red shift, 12, 161
reductionism, 156
relational, 15, 23, 154
row, 51, 67–71, 78, 118

scale-constants, v, 30–32, 92–98, 109, 154–156
scattering, 22, 29, 32, 40, 47, 48, 55, 74, chapters 7, 8
self organization, 50
short-range forces, 122
signal, 64, chapter 6
(see also light-signal), chapter 9
similarity of position, 33, 132, 133
spin, chapter 8
standard model, 91, chapter 8
strong interaction, 122
subjectivity, 153
SU3 (special unitary group in three dimensions), 131

test function, 65
test-particle, 10, 22
testing, 63, chapter 6
theory-language, 3, 9–11, 19, 20, 124, 135, 136, 139, 147, 149
three-dimensionality, 131, 142, 148, 151
3-event, 4-event, 54
TICK, 54
Tractatus Logico-philosophicus, 9
Turing machine, 50

uncertainty principle, 18, 23, 123, 125
Ur-intuition, 63

vector space, 51, chapters 5, 6
vertex scattering, 126

wave-function, 3, 137, 158
wave-function collapse, 158
weak interaction, 93
weak/electromagnetic unification, 121

yes/no symbol, or existence symbol, 35, 47